AF601872

Voluntary Agencies and Rural Development

THE AUTHOR

Dr. (Ms.) Seema Nikalje is working at Ankushrao Tope College Jalna (Former MSS College of arts Commerce & Science, Jalna) as Head & assistant professor in Public Administration and social health researcher since 20 years. Dr. Nikalje holds a doctorate in Voluntary Agencies and Rural Development in faculty of Social Sciences. She is teaching undergraduate and postgraduate students in the department of Public Administration, mentoring 8 Ph.D students. Author has been a reviewer for *International Journal for Quality in Health Care* and Springer *International Journal of Public Health.*

Dr. Ms. Nikalje has achieved several fellowships for postdoctoral research abroad in Australia, Hungary and recently completed a postdoc in US at Ohio State University. Her postdoctoral research over the nine years has focused broadly on public health, women health, community health workers and health of rural and marginalised population.

Dr. Nikalje has attended various state, national and international conferences, seminars and also presented her papers and lectures abroad. She has a modest publication to her credit. Ms Nikalje has also obtained Diploma in Theology from senate of Serampore College, University Calcutta.

Voluntary Agencies and Rural Development

Dr. (Ms.) Seema Nikalje

2017

Scholars World

A Division of

Astral International Pvt. Ltd.

New Delhi – 110 002

Cataloging in Publication Data--DK
Courtesy: D.K. Agencies (P) Ltd. <docinfo@dkagencies.com>

Nikalje, Seema, author.
Voluntary agencies and rural development / Dr. (Ms.) Seema Nikalje.
pages cm
Includes bibliographical references.
ISBN 9789387057883 (International Edn.)

1. Voluntarism--India--Maharashtra. 2. Voluntarism--India--Marathwada. 3. Rural development--India--Maharashtra. 4. Rural development--India--Marathwada. I. Title.

HN690.Z9V64 2017 DDC 361.37095479 23

Published by : **Scholars World**
A Division of
Astral International Pvt. Ltd.
– ISO 9001:2015 Certified Company –
4736/23, Ansari Road, Darya Ganj
New Delhi-110 002
Ph. 011-4354 9197, 2327 8134
E-mail: info@astralint.com
Website: www.astralint.com

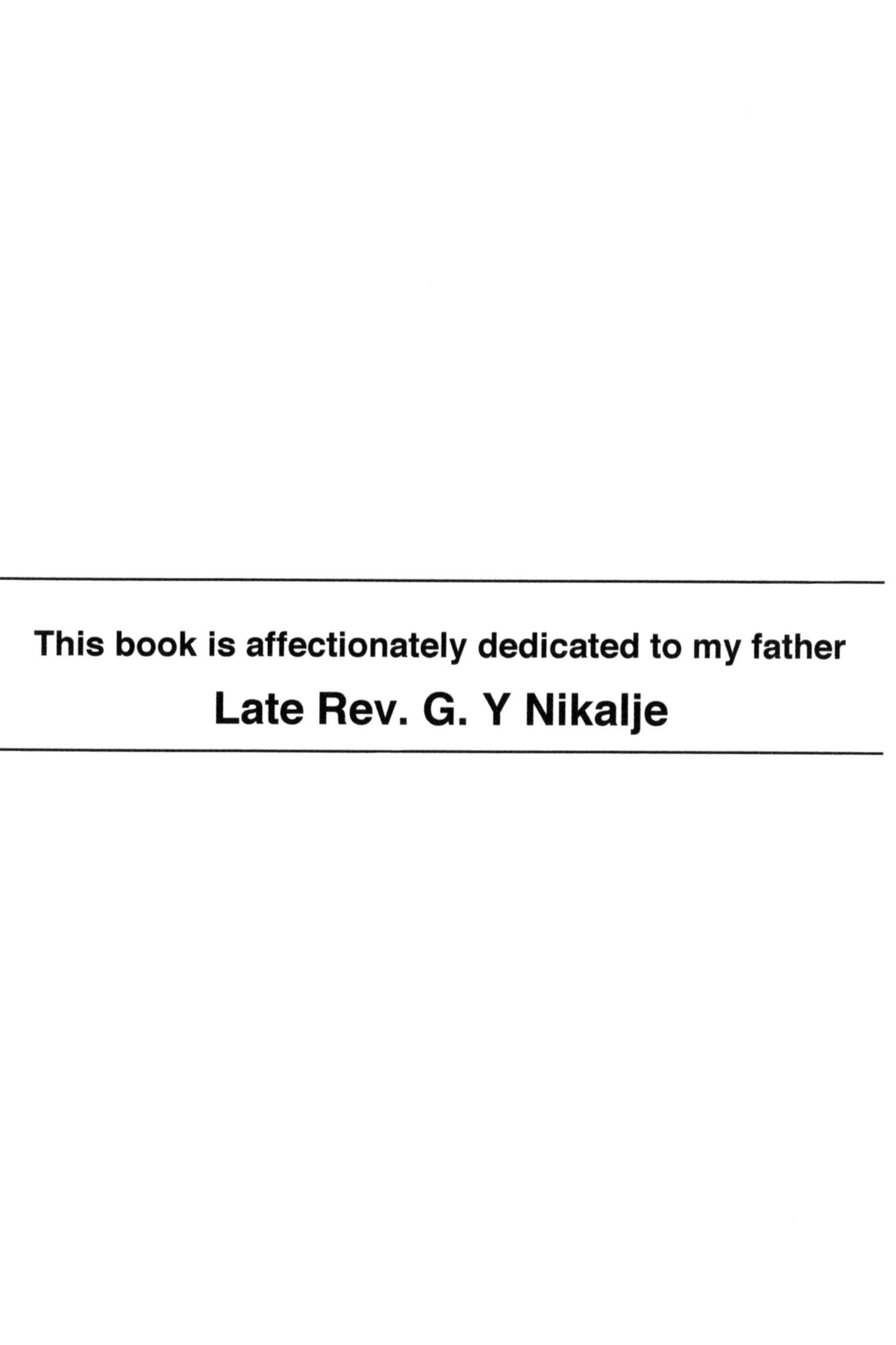

This book is affectionately dedicated to my father

Late Rev. G. Y Nikalje

Acknowledgment

I am indebted to Ankushrao Tope College Jalna, the President, late Hon. Ankushraoji Tope of the Matsyodari Shikshan Sanstha for his incessant encouragement and inspiration to pursue my higher studies.

I am thankful to the president of the institution, Hon. M.L.A Shri. Rajeshji Tope, Administrative officer Dr. B.R. Gaikwad and principal Dr. R.J. Gaikwad for supporting me in pursuation of this work.

I acknowledge with deep gratitude the help and guidance rendered to me by my mentor, Dr. P.M. Bora, superannuated Professor, Department of Public Administration, Dr. Babasaheb Ambedkar Marathwada University, Aurangabad.

I sincerely express my thankfulness to the National Institute of Rural Development library, Hyderabad. I thank the libraries of IHMP Pachod, KVK Jalna, JES College Jalna, Dr. Baba Saheb Ambedkar Marathawada University, Aurangabad and my college library.

I would be ever grateful to Marathwada Sheti Sahayya Mandal (MSSM) for giving me an insight into the watershed and rural development study. I am sincerely thankful to Shri. J.M. Gandhi, Director, MSSM for his valuable time and support. I am thankful to Shri. Vijayanna Borade, the nodal contact person, MSSM for his unending encouragement and help. I am also obliged to Shri Yugandhar Mandavkar, the then Project-in-Charge (Adgaon Project) for sharpening and honing my techniques and skills as researcher. I pay tribute and thank late Mr. Navshervanji Jalnawala, who consistently guided me into the working of MSSM.

I sincerely and humbly thank the people of village Adgaon and will remain grateful for their co-operation and truthful information.

I thank my former HOD, principal Dr. B.S. Pimple. I am obliged to each of my colleague especially Dr. Deore S.K, Gabhud S.D, Dr.Aseri A.A. for their support during this work.

I am grateful to all my friends for their encouragement and especially who accompanied me to the various voluntary agencies, libraries and the project areas.

I earnestly thank my friend who guided and helped tolerantly during my writing process. I also acknowledge with love to my family and well-wishers for their unfailing support and encouragement.

I will forever remain grateful to my dad, late Rev. G.Y. Nikalje, for his never ending encouragement for introducing this work in a book form for enlightenment of rural community. I also deeply acknowledge all prayer warriors and coworkers in prayer for supporting and praying for my every academic excellence.

I will always remain grateful to the Almighty God for giving me this opportunity and with it the timely help, confidence and wisdom needed for the completion of the study and now this book.

Dr. (Ms.) Seema Nikalje

Preface

I thank Almighty God for bringing this moment of writing the preface for my thesis to be converted into book, I have been praying for this moment since a long time.

Let me mention that in priority that this book will be a valuable tool for the rural community particularly the small scale farmers and NGOs to adapt and learn from the watershed case study that has been described in the book. I wish that I may perhaps translate this book in Marathi and Hindi languages which can be comprehensible for many in this region.

India truly has a very long history of Voluntarism and during this research an effort was made to study in depth the pros and cons of voluntary agencies in India and also to make a micro study of few Voluntary Agencies in Marathwada region of Maharashtra. Especially the Adgaon Watershed Project of Marathwada Sheti Sahayya Mandal (MSSM) has been distinctively studied during this research. We should be aware that Marthawada falls in the drought prone zone of Maharashtra with acute scarcity of water and lowest Human Development Index in a couple of districts.

During recent years there has been a growing recognition of the significant role that Voluntary Agencies, also known as Non Governmental Organizations (NGOs), are playing in the development process of our country, especially the rural areas. This has happened particularly for bringing about decentralized progress with the focus on promotion of the socio-economically backward regions of the country. Government has also realized the immense prospective of NGOs in the development process as can be seen from the approach adopted since inaction of several Five Year Plans. The study of NGOs in rural development has therefore become increasingly important particularly as there is paucity of NGO literature in our region.

The present volume is an attempt to meet the above felt need and to throw considerable light on the nature, role and responsibilities of NGOs as a promoter of development, their comparative advantages and limitations vis a vis the role of the government, evolution of government policies towards NGOs and their partnership, and contribution made by specific NGOs in promoting the overall comprehensive development of rural areas in the backward regions of the country.

The chapters in the book study the definition, origin, theories and sources of voluntarism in India. It also studies the seven phases of growth of voluntary agencies giving some vital typologies. It includes the history of Voluntary Agencies in India, Maharashtra and Marathwada and portrays the tremendous role of Voluntary Agencies in bringing about the rural development in India. The Adgaon watershed project has been critically studied along with a few Voluntary Agencies in Marathwada. Themes like Peoples' participation in rural development through voluntary agencies and empowerment of rural women through Voluntary agencies are also explicitly written.

An effort has been made to study the evolution of voluntary activities in State of Maharashtra and Marathawada region.

The volume would be useful to the students of public administration, rural development, researchers, social scientists, bureaucrats, policy makers, functionaries of Panchayat Raj Institutions, farmers, NGOs and all those interested in socio-economic development. The beneficiaries will also find it useful as it makes them aware of the various advantages a NGO can render.

Dr. (Ms.) Seema Nikalje

Contents

Introduction

Development is a comprehensive term and multidisciplinary in nature. Development may be defined as "an induced and properly guided positive change in a given situation at some place over a period of time". This may be said a process of desirable change in entire social, economic, technological, cultural, psychological, organisational and political aspects of human society.

Concept of development

Development as the term signifies growth or change for the betterment in all aspect of the social process. It can be related to the peoples' development or the development of the areas. Applied to an economy, it means growth in its different sectors of production and distribution, improving standards of education, living and civilization of the people, improving wages for the workers etc. Social development signifies the development of social institutions which may facilitate smooth changes in the outlook of people .It means improvement in social structures like norms of mutual conduct among members, values, humanization and modernization. "Development is a process of acquiring a sustained growth of a systems capability to cope with new, continuous change towards the achievement of political, economic and social objectives."

The most important component of development is its dynamism. Since it is a continuous process the existence of certain prerequisites has to be ensured viz: a stable political system; stable administrative system; a system of planning; and most importantly peoples participation.

The objectives of the development are;

- A substantial increase in living standards that encompass material consumption, education, health and environment.
- An increase in the capabilities and opportunities of their use by all people.
- Fulfillment of basic need to lead a self-respected decent life from all forms of servitude and exploitation.
- Removal of poverty and disparities.

- Increase in cultural, social and human values.
- Achievement of the ultimate end-human development.

Rural Development

Rural Development is yet to be given a acceptable definition. It has two words, rural and development. Rural means a rural society or specific areas marked off by villages which are lying outside the jurisdiction of municipal corporation or committees and not classified as urban areas.

A rural society has a socio-economic structure formed of different components inter-related with each other through their respective functions which are different from urban society. Social structure of a rural society consists of institutions like social stratification based on caste, class, religion, family, community status, power domination and variety of social contracts. The economic structure consists mainly of economic activities concerned with the system of production and consumption that are closely inter-linked with the social structure.

Rural development has been used in different forms such as a concept, a phenomenon, a strategy and a process. As a "concept"- rural development has been used in a comprehensive and multidimensional form involving the development of all rural socio-cultural and economic activities, infrastructure facilities, physical and human resources with a view to improve the quality of all rural people. As a "phenomenon" - rural development is the result of interactions between various physical, technological, economic, socio-cultural, institutional and environmental factors in the rural areas. As a "strategy" rural development is designed to improve the economic and social life of a specific group of people - the rural poor.

It involves extending benefits of development to the poorest of the poor among those who seek livelihood in the rural areas. In other words rural development strategy may be defined as a "process" of the transformation of a traditionally stagnant immobile and underdeveloped rural society into a modern one through the diffusion of knowledge, skills, values and technology, etc. In a nutshell the "process" of rural development must represent the entire gamut of change by which a social system moves away from a state of life perceived as unsatisfactory towards a condition of life regarded as materially and spiritually better".

Rural development acquired special significance as almost 70 percent of worlds poorest live in rural areas of third world countries. Rural development basically reflect four factors of development :

- Economic
- Social
- Political and
- Cultural

Hence development no longer means agricultural development or socio-cultural development, though it is a major facet as rural economy is based on agriculture. The nature and scope of rural development is very wide and its implications are politically and socially far reaching. The aim of implementing various rural development programmes is to improve the living conditions of the rural poor. In the new perspective, rural development no longer means agricultural development alone. It is not a social welfare case of pumping money into rural areas to provide for basic human needs. It encompasses a spectrum of activities and human mobilization to make people stand on their own feet and break away from all the structural disabilities which can chain them to the condition in which they live. Therefore rural development can be viewed as the self sustaining improvement of living conditions of rural inhabitants. Hence the objective of rural development is to enable all rural people to realize their full capabilities, thereby, fostering a better quality of life.

It also implies to make rural areas more productive and less vulnerable to natural hazards, poverty and exploitation and to give them mutually beneficial relationship with other parts of the regional, national and international economy. It attempts to ensure that any development is self-sustaining and people'sparticipation in the development process ensures as much local autonomy and as little disruption to traditional custom as possible. Participation in decision making process especially decisions affecting them is also an aspect of development. This is attempted to be achieved through decentralized administration - Panchayati Raj.

In India today, there is a strong emphasis on development of rural sector. A sizeable number of organizations have been involved in welfare and development works in rural areas. Inspite of attempts by a number of organizations to develop the rural sector like the State and Central government, financial institutions, industrial houses and voluntary organization, success has remained uneven and limited due to various reasons. In various fields voluntary organizations have played a vital role in social progress and have taken an active involvement in the promotion of human welfare and well being.

Development of village industries, agriculture and allied activities through technological innovations are the common activities of a large number of organizations engaged in rural development. The voluntary organization initiated development in rural areas in specific localities. There are thousands of small voluntary groups, unknown and invisible who cumulatively constitute a force which builds a healthier society. They contribute to social transformation through their modest efforts at organizing help for the needy, address their grievances and fight against atrocities.

NGO's are today the efficient harbingers of change in the society. India has a great tradition in voluntary work and the NGO's have played a significant role in rural development. In the Indian context the rural development refers to the process of improving living conditions, providing minimum needs, increasing productivity and employment opportunities and developing potentials of rural resources.

The causing factors for the failure of rural development programmes in India initiated by the government were the non-involvement and absence of people for whom the programme was meant. The need for micro-level institutional arrangement to involve people in formulation, implementation and monitoring of the programme is stressed upon to make the programme viable. The non-governmental organization (NGO) by virtue of being small scale, flexible, innovative and participatory are more successful in reaching the rural poor. The reach of voluntary action in rural society is now expanding in tune with the observed needs and scope. Voluntary agencies are credible in the sense that they motivate the rural masses for a change of attitude and perception. They also generate awareness about the rural development programmes on the one hand and alternative paths on the other. They also through repeated reinforcement induce refinement into the programmes.

The word NGO is used by for organization functioning outside the government sector. According to Bunker Roy trade unions, co-operatives, political parties, khadi boards etc. are considered NGO's but not voluntary organization. For voluntary action we need organizations which are indigeneous, decentralized and non-statutory with direct participation of the beneficiaries or target groups in the programme. Voluntary action is now concieved in a broader and combined connotation as an admixture of voluntarism and activism.

Organized forms of popular participation make development stable. Participation inculcates a feeling of community belonging and thereby initializes the whole process. The Planning Commission, therefore stated that, " properly organized voluntary efforts may go a long way towards augmenting the facilities available to the community for helping the weakest and the most needy to a somewhat better life." To supplement these needs of development, the Planning Commission recognized the role of NGO's in the first five year plan document itself in the field of social welfare, and the third plan characterized the voluntary action as an aspect of public cooperation. The Planning Commission made a beginning in this direction in 1982 itself, when the then Prime Minister, in a letter to the chief ministers of the States emphasized the need for widening the role of NGO's in implementing the 20-point programme with a suggestion to set up a consultative group of NGO's under the chairmanship of development commissioner. Participation of voluntary association in the process of decision making and decision implementing is necessary. The seventh plan (1983-90) looked at voluntary organization as "the eyes and ears of the beneficiaries" and the weaker sections of the society who have been left out of the development benefits. As a strategy voluntary associations would be closely and critically associated to the target groups so that they receive development benefits meant for them, augment the programmes and also set up a feedback mechanism for these associations.

The variety of voluntary association can be placed on a developmental axis considering the methods of development, contents of programmes and basic objectives guiding developmental processes. The lowest scale would be " duty bound individual supported by a religious frame and on the highest would be citizens groups activities supported by a secular frame.

The voluntary agencies have been involved in rural development much before government's effort in this field. They were the early catalyst for new thrust in rural development and their role was that of pioneer, innovator and scientist. The voluntary organisations have expertise and capabilities to work in many non-traditional areas and thus implementing various rural development programmes.

Need for the study

Drought is a climatic phenomenon characterized by prolonged periods of low humidity, weather and scarcity of rain. Rural communities dwelling in drought-prone regions lead miserable lives. Drought affects cultivation of crops and causes situations leading to famine. Various countries located in different continents of the world face famine and scarcity. Especially India, falling within the monsoon type climatic region, with its large percentage of population depending on agriculture and a history of fluctuating rainfall, faces famine and scarcity very frequently. A considerable portion of the Indian population lives in drought-prone areas.

In Maharastra, drought is an old and recurring ailment, affecting one third of its area and about 37% of its rural population. The Fact Finding Committee for survey of scarcity areas in Maharastra state 1973, listed the effects of famines in Maharastra state as follows ; failure of rains/drought; withering of crops; burning of grass; reduction in drinking water supply; rise in price of grains; increased demand for employment; starvation deaths and deaths from epidemics; loss of cattle due to fodder shortage; migration of cattle; sale of ornaments; thefts of grass; emigration of persons in search of employment and increase in mortality. The study has undertaken the work of Adgaon Watershed Development Project, which was carried out by the regional voluntary agency - Marathwada Sheti Sahayya Mandal (MSSM). MSSM is a voluntary organisation working in central Maharashtra for rural development.

Gravity of the Problem

Marathwada lies in the drought belt of Maharashtra. It is backword region in Maharashtra. The agronomy is dependent proportionally to the water resources available. The socio-economic, educational, industrial, health, women development are also directly related to the degree of agricultural development. The development of this region is also lagging behind because of the apathy and negligence of the government and the political bodies.

Research Methodology

The case study approach has been adopted for this research to achieve the objectives. Primary data was colected by several visits to the Adgaon project areas and some important NGOs working in Marathwada - Marathwada Sheti Sahayya Mandal (MSSM), Aurangabad, Manavlok (Ambejogai), Abhinav Vikas Sanstha (Fardapur), Grasp (Aurangabad), Jannarth (Aurangabad) Dilasa, Aurangabad, Institute of Health Management Pachod (IHMP), Comprehensive Rural Health Project (CRHP), Jamkhed and a few more, to observe the situation prevalent in the area and surrounding areas, including the work implemented. The visits to project enabled to have better insight and qualitative information by observing the discussions that leaders/executives of these projects had with trainee officers. Discussions and informal interviews were held with leaders/executives of the concerned voluntary and government agencies involved in the project. Informal interviews were held with some of the experts who had studied this project earlier. Discussions were held with the academicians, international activists and technical experts in watershed and government officials related to rural development. The secondary data was obtained from various research papers, reports and journalistic articles published and unpublished and content analysis was done in order to extract relevant information contained in these material.

Aims and Objectives

- To trace the history, evolution and development of NGOs in India, Maharashtra and Marathwada.
- To study the roles of NGOs in rural development.
- To study the impact of the various NGOs operating in Marathwada with reference to their organisation principles, methods, techniques and activities.
- To highlight the role of MSSM in Adgoan Project with reference to its development projects in drought-prone areas. To study the impact of socio-economical, ecological, technological and administrative and women empowerment and people's participation in the process of development.
- To study the programmes and policies of government organisations and NGOs in rural development processes and to emphasis the collaboration between them.
- To propose the guidelines for future successful collaboration and joint working of NGOs and GOs for rural development.

Scope of study

Based on the above, the study proposes to put forth the guiding principles effective and sustaining for the NGOs and the government for rural development. It will also be helpful for the government to bring reforms in policies, approaches towards NGOs .The study will benefit the farmers and villagers for implementing systematic watershed projects for rural development.

Area of the study

Area of the study undertaken was the village of Adgaon in Aurangabad district of Maharashtra. MSSM is working there since 1984. The study also consists the evaluation of random voluntary agencies which are working in the other district of Marathwada. This was to explore several areas of NGO involvement and the impact they have on the rural community.

Hypothesis

- In the region of Marathwada despite of government programmes implemented for several years after the Marathwada freedom struggle there are no tangible results observed.
- The poor development of Marathwada is due to its location in the drought prone region, lack of underground water reserve, migration of farmers for jobs into cities and poor educational standards.
- Irrigation system of Marathwada was mostly rainfed and until 1956 there was no surface irrigation. Recently the government has small irrigation

projects but sufficient to maintain Rabi and Kharif crops. People have no or little awareness of rainwater harvesting and watershed concept.

- The failure of voluntary agencies to work effectively in Marathwada was because of poor objectives and poor participation of the local people.

Literature Review

Rohini Patel illustrates about Voluntary Agencies in India: Their Motivations and Roles in her writing. The concept of the third sector and the variety of factors motivating and shaping the strategy of voluntary agencies are dealt with. The theories of origin, viz., Contract Failure Theory, Economic Theory, Subsidy Theory, Exchange Theory are utilized in the thesis.

Rajini Kothari analyses the reasons why, despite best schemes for development, we have, infact, displaced the poor in the name of development and have allowed trends of pro-elitist bais and dependence on the Central government instead of building social organisations and social base over the years. Kothari pleads in the name of voluntary agencies the case of "activist and non- party, non-state, groups and organisations" to extricate us from the present morass of crises.

Muttalib discusses the various concepts, with their behavioral overtones, involved in voluntarism, including their typologies. He is however optimistic about the future of these bodies as according to him, "societies with a heritage of authoritarian forms of social and political institutions, inhibit the growth of voluntary organisations."

Bunker Roy discusses the shift that has come about in government thinking regarding involvement of voluntary agencies in development since the sixth plan period. He then delineates the role of these bodies as visualised by the Planning commission and the programmes in which they are to be involved. Roy also discusses briefly the new criteria for identification of these bodies for enlisting development programmes and a code of conduct , drafted by Planning Commission for them.

Nandedkar, treating development more as a social phenomenon in view of the social context, endorses the Gandhian model of participative association of people in development owing to the negative orientation of governmental bureaucracy. He discusses different types of voluntary bodies which can possibly be involved in development programmes.

D. Rajsekhar in his NGOs in India: Opportunity and Challenges of Voluntary Agencies in India, traces the history and evolution of NGOs. He states the types of voluntary agencies, viz. Operational or Grass root NGOs, Support NGOs, Umbrella or Network NGOs and Funding NGOs. He further discusses the advantages and weaknesses of NGOs. He enumerates the characteristics of good NGOs.

T.N.Dhar in his NGOs: Their Roles, Responsibilities and Problems have suggested some initiative for NGOs *viz.,* democratic decentralization and empowerment, empowerment of women and rural development and technology.

Alka Srivastava in her NGOs and Rural Development has discussed the concept and emergence of NGOs, areas of NGOs involvement in rural development and government and NGOs interaction in rural development.

Hoshiyar Singh in his Rural Development Concept, Strategy and Approaches discusses the different strategies and approaches to rural development, viz., the multipurpose approach, minimum package approach, target group approach, area development approach, spatial planning approach and integrated rural development approach.

M .Laxmi Narasaiah and G. Jaya Raju in their Rural development and Anti- Poverty Programme discussed the need for rural development and various government anti-poverty programmes, viz., SFDA, MFALDA, CSRE, FWP, MNP, ADP, DPAP, IRDP, DDP etc.

Shakuntala Narasimhan's work spotlights on rural schedule caste and schedule tribe women who are triply disadvantaged- as women, as rurals and as dalits and tribals in her Empowering Women an Alternative Strategy from Rural India. She has assessed the effectiveness of an alternative strategy of development and empowerment of women that begins with awareness generation and motivation rather than economic interventions.

S. Nagendra Ambedkar in his People's Participation in Rural Development discusses the need for people's participation, bureaucracy and participation. C. Francis in his Rural Development: People's Participation and The Role of NGOs gives the strategies for NGOs to employ and enlist people's participation.

Ramesh Vasvani in his book Micro-Watershed Development: Three success Stories from Maharashtra have done a detailed analysis of the Adgaon project highlighting the pre-project and post-project conditions. He emphasises the principles of leadership, people's participation and the impact Adgaon project had on the village, viz., changes in cropping pattern and land use pattern, animal husbandry and dairy farming and social infrastructure.

Marathwada is the most backward among the backward areas of Maharashtra because of neglect by the State and political powers. Dr. Mrs Kaldate in her study of the social change in Marathwada opines, the backwardness of Marathwada is because of lagging behind in four major directions - population control, literacy, irrigation and social and educational services. These inadequacies have gone on increasing because of the lack of public participation and lack of mobility on the part of social and political leaders while observing the changing old habits and traditions in a long process, the authoress observes that voluntary organisations are striving to bring about the change and have achieved success in overall development but caste, religion and son of the soil considerations are sometimes used as weapons to strike out at these successes.

Dr. Kurulkar's study on the economic development of Marathwada states that the co-operative movement and development programmes are weak and ineffective because of the educational backwardness of the region and due to the remnants of the feudal and semifeudal social system of this area.

There is a paucity of literature on rural development programmes, strategies and its outcome in the Marathwada region. There are a few voluntary organisations working in selected sectors without much publicity in narrow confines, their asset is accumulated experience and faith of the people rather than macro planned project baskets.

Yugandhar Mandavkar conducted a detailed socio-economic impact assessment studyof Adgoan Project in 1991-1992. The results showed that the project had achieved considerable equity across classes, with the landless and small marginal farmers benefiting for more proportionality than the erstwhile "better-off" farmers.

Attempts were made to study equity on social aspects of the project, prominent of them was the gender analysis study conducted in 1993-1994 by Dr. Uma Ramaswamy and Mrs. Bhanumathy Vasudevan. This study was undertaken to understand the impact of the watershed development on women in Adgaon. This study brought out some striking disparities among men and women with respect to the benefits, work load, technology and options generated by the project.

The few reports on work done by VOs in the Marathwada region have been reviewed. Reports by S.C. Jain, S.K. Verma, Manisha Kale, Nitin Inamdar in the participatory watershed development programme report (1996-2002) states that the watershed programmes have improved the economic status and overall development of the villages with promising local participatory effort.

This study of MSSM with reference to Adgaon project was taken up because of its uniqueness. The Adgaon project proved to be a pilot watershed project where concepts of participatory natural resources management with watershed as a unit was initiated. Adgaon project is a national example of integrated watershed development. This project was helpful to development planners and practitioners to formulate policies and programmes for equitable development of poverty alleviation and drought proofing on a sustainable basis.

There is no comprehensive study or review of this project published. The paucity of literature and documentation has made this study very essential though cumbersome and painstaking.

The study of this project will be the first of its kind. This project in Adgaon has demonstrated to a pragmatic, viable and replicable approach to management of land and water resources on the watershed basis to overcome the ill effects of drought. Experiences of MSSM projects indicate that the community have not only ably shouldered the responsibility of maintaining and managing the resources created, but have gone beyond to undertake further development initiatives on their own.

The thesis is divided in the following chapters :

Chapter -I

In this chapter an effort has been made to trace the origin and development of NGOs in India, Maharashtra and Marathwada. This chapter also deals with the concept of voluntarism.

Chapter-II

The chapter focuses on the concept of rural development and defination of NGOs. This chapter highlights the role of NGOs in rural development, and the typologies, the components of rural development, the selected rural development programmes in the country and the State. It also studies disaster relief programmes. The chapter studies advantages and weakness of NGOs in rural development and its future challenges.

Chapter-III

It deals with some major NGOs like MSSM, GRASP, Dilasa, Jannarth, Abhinav Vikas Sanstha, Nirman and Manavlok of Marathwada, with reference to their organisation principles, objectives, methods and activities. It also depicts the genesis and development activities of the NGO MSSM in particular.

Chapter-IV

This chapter is a case study of Adgaon project. Adgaon's history and geographic location is also mentioned. It studies the evolution of the project, the philosophy of watershed, the pre-developmental assessment and the impact of the project in various socio-economic technological and administrative sector.

Chapter-V

This chapter contains the need for people's participation for rural development, many different programmes where successful participation has been sought and the strategic elements for successful participation for rural development.

Chapter-VI

This chapter deals with the rationale for women's empowerment conceptual framework. Processes of development, State initiatives towards women development and critical evaluation of various women programmes and women potential in watershed development is described.

Chapter-VII

The crucial role of NGOs and GOs in development has been studied during various Five Years Plans. The chapter also focuses on the emphasis of the role of PRIs in rural development with the cooperation of NGOs. It also depicts different initiative proposals of NGO and GO collaboration.

Chapter-VIII

This chapter contains the conclusion and future guidelines including successful NGO-GO collaboration in rural development. It also contains useful suggestion for better functioning, policy making, administration and governance of NGO, GOs in rural development.

References

I.S. Hooda : 'Rural Development' Some Conceptual Issues; Journal of Rural Reconstruction, 29(1) 1996, P. 1.

World Bank (1990) World Development Report, Washington D.C.

Madhu Mishra : Rural Development-Action and Management (ed) P.61-65.

Readings on Rural Development Programmes; IAS, Professional Course (Phase-1) 1993-94.

Seventh Five Year Plan 1985-90, Vol II, Delhi, Government of India, Planning Commission, 1985.

Dr. Mrs. Sudha Tai Kaldate; Social Change in Marathwada : Development of Marathwada; A Perspective; Publication, Swami Ramanand Teerth Research Institute, 1999. P. IX.

Dr. R.P. Kurulkar, Economic Development of Marathwada : Development of Marathwada; A Perspective; Publication, Swami Ramanand Teerth Research Institute, 1999. P. 38.

Yugandhar Mandavkar, Equity in Watershed Development, Paper Presented at The Workshop on Participatory Watershed Development Organised by Swiss Agency per Development and Co-operations (SDC) Project.

S.C. Jain, S.K. Verma, Manisha Khale, Nitin Inamdar : Final Evaluation Report, Participatory Watershed Development Programme (1996-2002).

CHAPTER - 1

History of Voluntary Agencies in India, Maharashtra and Marathwada

Introduction

This chapter studies the definition, origin, theories and sources of voluntarism in India. It also studies the seven phases of growth of voluntary agencies giving some vital typologies. An effort has been made to study the evolution of voluntary activities in Maharashtra and Marathwada.

India has a long history and tradition of voluntary action, providing services to the sick, needy and destitute. Rather, it is a part of our cultural heritage and way of life. Voluntarism in India is as old as the emergence of organised society itself. It originated as pure philanthropy of charity and this motivation sustained the voluntary efforts all through history in the ancient and medieval period. The voluntary efforts in the process of welfare and development have undergone evolutionary changes with changing emphasis on various experimental development programmes in India. The history of voluntary action is an integral part of the study of evolution and changes in the Indian society.

Definition of Voluntary Agencies

The term voluntary association is variously defined. Michael Banton, an anthropologist, characterized it as a group organised for the pursuit of one interest or of several interests in common. Usually, it is contrasted with voluntary groups serving a greater variety of ends, such as kin groups, castes, social classes and communities.

David L. Sills, a sociologist, identified it as a group of persons, organised on the basis of voluntary membership without state control, for the furtherance of some common interest of its members. A more recent development is the government-developed and - mandated programs seeking increasing citizen participation - specially the poor - through creation of necessary, local structures although some viewed them as contradictory to the autonomy of voluntary associations. Sills included three types of similar associations: 1) Making - a - living association like business firms, trade associations, production, marketing and consumer cooperatives, professional associations and labour unions; 2) religious organisations; and 3) Political parties. He contented that membership in such voluntary associations as labour unions or professional societies, may be a condition of employment or professional factor and, thus may not be truly voluntary, membership in church or in a family may be inherited from one's parents and in that sense, not voluntary nor they are drafted into them as in the case with military, as observed by Constancy Smith and Ann Freedman.

Bourdillon approved of these inclusions for different reasons, more specifically articulated. He observed that religious bodies, primary purpose is glory and service to God; political parties and organisations primary purpose is to gain power; and organisations which may make provision for social well-being, but actually may have primary economic motives behind them. He also included two other types of organisations - one based strictly on insurance policies, where payments are made by members in consideration of certain specific benefits to be drawn by them; and the other pursuing art for arts sake.

Norman Johnson, who examined at length in his recent study the various definitions of voluntary social services, found their turning on four factors: 1) Method of formation, which is voluntary on the part of a group of people; 2) Method of government, with self-governing organisation to decide on its constitution, its servicing, its policy and its clients; 3) Method of financing, with at least some of its revenues drawn from voluntary sources and 4) motion with the pursuit of profit excluded.

Legal status is another debatable issue receiving serious attention. Sills held that whether a voluntary association is incorporated or not, it has few consequences for its activities. Nevertheless, in the Indian content, the distinction has assumed importance for their financial accountability. Now it is clearly suggested that only those voluntary associations would be considered for grant-in-aid which are incorporated and have been existing for at least three years. If the former condition provides a legal basis of relationship, the latter assures some amount of organisational stability. Both will help secure formalization and enforce financial accountability.

Smith and Freedman considered voluntary association as a structure formally organised, relatively permanent, secondary grouping as opposed to less structured, ephemeral or primary grouping. Formal organisation, they said, is identified by the presence of offices which are filled through some established procedures, schedule meetings, qualifying criteria for membership and some formalized division and specialization of labour, although the organisations do not necessarily exhibit all these characteristics to the same degree.

With the increasing involvement of voluntary organisations in welfare development, they have to sacrifice their autonomy. There are quite a few restrictions which voluntary associations have to accept if they expect public grants. These are regulatory characters.

In India, for instance, religion besides politics, is the other social sphere from which they have to keep themselves away if they wish to seek public money for participation in nation - building activities. This is in consonance with Indian secularism, which prohibits use of public money for propagation of any religion.

Infact, besides the above-mentioned specific conditions, a more comprehensive framework is determined for them. Thus like the statutory agencies, they must be committed to national objectives, namely, socialism, secularism, democracy, national unity and integrity. Although they may restrict their autonomy, they help smoothen their freedom without national obligations.

Voluntary organizations in India are shaped by a variety of factors; they have been nurtured and threatened in turn as the dominant political outlook has oscillated between liberalism and statism. Diverse in their functions and purposes, they serve primarily in vast and assorted population of informal workers. A sizable sum from public and foreign sources is expended by this sector. And yet the sector has received surprisingly scanty attention from academic researchers and social analysts.

As the voluntary agencies are diverse they are distinct from both public government and private organizations and, accordingly, should be brought under a common conceptual umbrella. Since voluntary agencies do not exhibit characteristics of either public or private sector organizations, the concept of yet another sector-variously called the third sector, independent sector, voluntary sector, or non-profit sector- is necessary to encompass all voluntary agencies.

Sources of Voluntarism

The term voluntarism is derived from Latin word Voluntas' which means Will'. The will assumes various forms of impulses, passions, appetites or desires. It is prior to or superior to the intellect or reason. All theories of voluntarism, whether psychological, ethical, theological or metaphysical which interpret various aspects of experience and native in the light of the concept of the will, subscribe to the thesis. It is the will that may produce miracles' - and thereby, some of the social evils, of which the unfortunate sections of the society are the victims, can be eradicated.

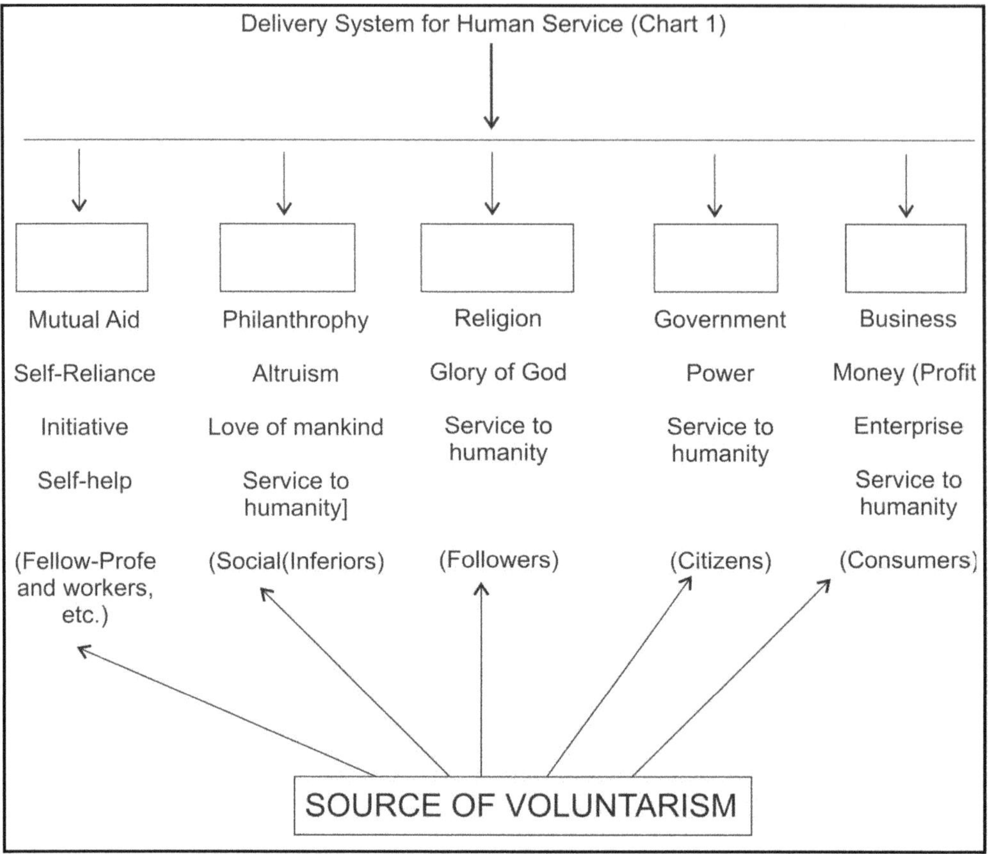

All voluntary associations which are the expression of human impulse] will , have been the subject of study of, by and large through disciplines, sociology, social psychology and public administration, sociologists, whose contribution is significant in this regard, study the associations as part of the social system, social psychologists are concerned with voluntary associations in an environment of their individual members; and the students of public administration with their organisational processes.

No comprehensive research effort is so far made to explore the motivating sources of voluntarism. Inequality among individuals is a perennial source of motivation for voluntary action in any society, whatever be the stage of its development. Inequality may be in material, moral or intellectual terms. A.F.C. Bourdillon observed that there is always an element of betterment of compensating inequality-motivating individuals to volunteer their services. Social services are designed to help compensate for the inequality and to uplift the under-privileged. The minimum standard, which the community can tolerate, he added, for its citizens, is not static. It moves forward from one sphere of activity to another.

The term volunteer' is normally used to denote someone who offers unpaid services to a good cause. Bourdillon remarks that every voluntary organisation is a product of the blood, sweat and toil of a few individuals, who were known for their persistent efforts for achievement of their sincere aspirations.

One may identify five sources of voluntarism (shown in Chart 1) , religion, government, business, philanthropy and mutual aid if the delivery system for human services is analyzed.

The missionary zeal of religious organisations, the commitment of government organisations to the public interest the profit-making urge in business, the altruism of the social' superiors and the motive of self-help among fellowmen; all reflect in voluntarism. At the operational level, the above mentioned components may not differ much from one another but each of them is moved by an impulse with services as the common motivation.

Bourdillon and Williaim Beneridge viewed mutual aid and philanthropy as the two main sources from which voluntary social service organisations would have developed. They spring from individual and social conscience, respectively. Bourdillon observed that the schemes based on these two motivating factors, are converging on each other and it is this mixture of the two elements which is peculiar character of the voluntary organisations today.

G.D.H. Cole identified the problem of poverty as the point of focus in all types of associations. One essential element is that of philanthropy, pure and simple, which has been always present, but with it has been mingled, from time to time, motives and impulses which have changed greatly from age to age, attributing the sufferings of the poor people either to religious or secular roots. Some of the most ardent social reformers of the 18th century and early 19th century, who wanted to do many things., were insistent that it must keep its hand off education. They drew the line between physical welfare, which the State could legitimately arrange to promote, and intellectual and moral welfare, which was a matter of outside the scope of political action.

William Beveridge put forward the idea of expansion of the social welfare by enunciating the concept of social insurance against the evils of want, disease, ignorance and squalor and the citizens entitlement to welfare as of right, a kind of social security against misfortunes.

Cole believed that although religion and social service have been most intimately intervened in the early stage of voluntary organisations, there have been philanthropists at all times, who were not moved by religious motives. Along with changing impulses, which removed the upper class philanthropists in doing good to their social inferiors with the welfare state , these inferiors began to play a significant part in looking after those and creating associations of their own, instead of receiving passively and as individuals, the benefaction of the well-to-do.

With the acceptance of the obligation of the State to ensure for all its citizen a basic minimum standard of life, the State emerged as the major philanthropist. But the voluntarists and advocates of the right of the needy to be maintained at the hands of the State, were looked with suspicion in everything under their auspices.

Nevertheless, it soon proved to be a passing phase in the developed countrieswith the realization of their mutual roles. Now there is no reason, as Cole remarked to suppose that as the scope of State action expands, the scope of voluntary social services necessarily contracts. On the other hand, its character changes in conformity, both with the changing views of the problem of State action and with the growth of the spirit and substance of democracy. Indeed, as he remarked, as long as there are rich and poor, a social service will necessarily continue in some degree and reflect inequalities of class and income.

Theories of Origin

To explain the genesis of voluntary organizations in particular societies at specific time periods, theories of the origins of voluntary associations help explain this phenomenon on a different plane.

Western scholars have put forward several theories regarding the origins of the voluntary sector. The disciplines from which they emanate range from economics and political economy to sociology, political science and anthropology. The theories are partial in that they are not incompatible with one another.

The Contract Failure Theory, an economic theory, explains why particular kinds of goods are produced by voluntary rather than by the private sector. It argues that when consumers feel unable to evaluate accurately the adequacy or quality of the goods, they choose voluntary organizations as suppliers rather than profit-making firms. Consumers distrust for-profit firms for such goods because the manager of a for-profit firm may supply inferior quality goods and pocket the additional earnings thus made. The opportunity to do so does not exist for managers of non-profit concerns as they are forbidden by law from garnering the extra profit. An enlightened consumer thus protects his interests by sponsoring voluntary associations.

The theory as it is formulated does not appear applicable to the Indian situation since in India most voluntary organizations are set up to meet the needs of the vulnerable and those who cannot protect their interests, and to deliver merit goods. A modern welfare state is expected to provide them, but India has failed to do so. The reasons for failure are many: First, the State lacks resources. Therefore, voluntary associations frequently supplement the supply of such goods, e.g., health care. Second, often, negligent public servants fail to perform their duty. Third, the weak and the vulnerable often do not know how to access merit goods; sometimes they even have to be convinced of their benefits.

The second economic theory, the Subsidy Theory, argues that non-profit organizations benefit from a variety of implicit and explicit subsidies, such as exemption from taxes. Thus, once set up, part of the financial burden shifts to the government, a prospect that acts as an incentive for setting up a voluntary agency.

The theory seems particularly applicable to the Indian situation since voluntary agencies here often obtain even their initial funds either from the government or foreign funding agencies. This theory explains the phenomenal growth in the number of voluntary agencies as a result of the abundant availability of funds, both foreign and domestic (mainly government), which began to characterize the voluntary scene from the late (1960)'s onwards.

The Exchange Theory offers yet another way of understanding voluntary agencies. It views a voluntary agency as a benefit exchange: the group organizer offers a set of benefits to the members and receives benefits in return. To join and continue as a member, one may have to pay a subscription, attend meetings, etc. These are the costs a member has to bear to receive the benefits. The organizer, on his part, has to devote time and energy to recruit members and to obtain and deliver benefits to retain them. The organizer's benefits may take the form of fulfillment of much-cherished goals or monetary compensation. The benefits that either party receives could be material, soldiery, or purposive. These theories lend a fresh perspective on voluntary agencies.

Development of Voluntary Agencies

The phenomenon of voluntarism is typical of the modern age, particularly since the industrial revolution in Europe during the nineteenth century and a little earlier. In the polities under the colonial rule of the European powers, the phenomenon of voluntarism appeared on the assurance of the capitalist economic processes, but markedly since their independence from the colonial rule. Though the immediate context of the relevance of voluntarism in development is modern, its social and cultural roots and moorings are traceable to the ancient and medieval times in the case of a country like India possessing a historical legacy of rich socio-cultural and economic achievements.

In Ancient and Medieval Times

In ancient and medieval India, voluntarism operated freely and extensively in the fields of education, medicine, cultural promotion, and even succour in crises like droughts, floods, epidemics and foreign invasions or deprovation by robbers and criminals. Pluralism played a great role in communication and transport network and the sway of the concept of dharma' obligation , Peoples' conception of functions of the state] meaning mostly the king was limited to maintenance of law and order and observance of good behaviour each unit of social organisation and government in the social and territorial hierarchy, viz., the individual, the family, the caste, the village, the temple, the guild, the town or city, the region and the highest level of government] mostly the king or emperor , owed certain responsibilities in development in the above restricted sense inherent in the conception of its own Dharma'. The ambit of Sasana' edicts of or directives of the king / emperor, was restricted by the rules of the Dharma Shastras, the guild, the caste and the local governments. A general statement of the historian that the ancient Indian State "sought to promote the moral, material, aesthetic and spiritual progress of the whole community" is to be construed within the above parameters of the social and political organisations in the country in ancient and medieval times. The poor and the destitute, the sick, the distressed and the disabled, the widows and the orphans, the jobless and the old, were provided succour by different social organs. The agriculturists, the artisans and craftsman, the traders and businessmen, the bankers and the professionals were helped out of difficulties by their colleagues in he occupation or profession. Philanthropy was widespread. The State came to the escue of the community in extreme contingencies of helplessness.

In India, before independence, in terms of scope of coverage in development activities, laissez faire occupied the largest portion, voluntary action through philanthropy, associational works and individual actions stood next, while State intervention came last. Here, development' excludes the infra-structural activity, like laying out of ports, railways and roads, setting up of communications in the form of posts and telegraphs or building up of essentials for capitalist development in banks, insurance companies, etc. Development' here includes instituting or extending facilities for agriculture, health and medicine, education and allied human development or enrichment aspects. After independence, State intervention claims the largest amount of development activity, while laissez faire would come next and voluntarism would have the lowest share of development activity. Voluntarism has receded into a negligible role in development because people in general in need of help, succour or assistance expect it to come from the government through its plan outlay. The scope of development planning has expanded more and more, successively, in each of the seven Five-Year plan had to caution explicitly that henceforward voluntarism had to come forward to play a greater role in development - rural and urban. The abolition of princely States dried up a source of philanthropy in development. The business and industry, another source of philanthropy in development, became reluctant to contribute funds for development due to alleged high taxation on them. The professions, the medicine, law and others, were exposed to increasing commercialization tending away from its role in development.

Kautilya's Arthashastra' might have depicted the expanded bureaucratic hierarchies of a central government of the Maurya Empire, indicating exceptional conditions of developed welfare-statism. The Gupta Empire also represented another existence of developed welfare-statism, though, incomparable in scope to the Mauryans Empire. The paragon of the Mauryan welfare-statism was missed from the government of Harsha, which followed a laissez faire policy.

The large scope of voluntarism in development during ancient and medieval India, however, did not mitigate the social inequities of the rigid caste-system, particularly heaped on the lower castes, especially the untouchables. The large-scale distress of and ravages on the people resulting from drought, floods, pestilences, robberies, crime, foreign invasions or excessive exploitation by the monarch, the aristocracy or government officials could not be saved by voluntarism, howsoever beneficent was its role in normal times.

In Muslims and British Era

The above delineation of the role of voluntarism in development in ancient and medieval India was not affected much by the inroads of the Muslims and the establishment of their rule in different parts of the country, including the Mughal Empire at Delhi and Agra. In Muslim society also, the above mentioned norms of voluntarism were applied. Barring the benevolent requires of rulers like Shershah and Akbar, the superimposition of another government, however, tended to entail heavier taxation and other levies on the peasants and other sections, which tended to depress their standard of life. Exemptions, concessions and moratoria were, however, granted in times of extraordinary difficulties.

During the British rule, voluntarism in development received a boost in new religious, cultural and social surroundings, though traditional sources of voluntarism lost the intensity. The laissez faire policy of the British government in economic, and apparently in religious and social matters left no other avenue of development open to the Natives' than resort to self-help, another term for voluntarism. The Christian Missionary educational institutions and hospitals and dispensaries set an example to the non-convert majorities both among Hindus and Muslims to emulate. Schools and colleges were established by educational societies set up by English-educated natives and affluent businessmen, traders, zamindars and members of aristocracy, to impart western education to the native children and youngsters. Pathshalas and Madrassas' imparting the instructions in traditional learning, however, suffered an eclipse. Libraries and lecture series were another source of social renaissance that ushered in cultural development of a novel kind by voluntary agencies. A new religious awakening was aroused by voluntary agencies stabilized in due course in institutional form. Some of these kindled generated sentiments of religious and social reforms. An awareness of the backward and suppressed conditions of the women, orphaned children and the backward castes and communities was roused among the educated by social and religious reformers and the associations founded by them. Later, self-awareness among these hapless sections was generated, which in due course evolved into associational and institutional activity. Educational institutions, hospitals, clinics and dispensaries and welfare associations that were set up also grew, which to an extent ameliorated conditions of these hapless sections of the society.

Voluntarism, thus, played a significant role in educational, health and medical and social welfare development in India during the later half of the nineteenth and the first half of 20th century. Not that the extent of development achieved was such as to solve the problems of development altogether, but a spirit of voluntarism was aroused, which gained a wide social recognition as an avenue of development during the British rule, before independence.

During the Days of National Movement

In contrast, with the growth of voluntarism in social and welfare development, new movements were not initiated in rural areas, including agricultural development except the cooperative credit movement in a few provinces, including Bombay and Madras. The original voluntary character of the cooperatives, however, was gradually overshadowed by the increasing State financial assistance to, and later underwriting of the credit cooperatives. In the industrial field, managing agency and other forms of legalized voluntary modes helped industrial and ancillary activity including banking and development in the country.

The national movement breathed a spirit of strengthening the normal fibre of the people through self-help and autonomy through independent institution building in education, industry, business and trade, and fostering of economic production, particularly of industrial goods through Swadeshi, i.e., own factories, workshops and crafts and Boycott of imported British goods. The triad creed of National Education, Swadeshi and boycott were advocated by Tilak, Aurobindo and others during 19th century to bolster Swarajya, the fourth component of the creed. Gandhi propagated the creed in the course of the non-cooperation movement of the early sufficiency was actively canvassed by Gandhi, and he lent it an institutional forms. Voluntarism, thus, secured a fresh lease in the national movement. Gandhi based it on the philosophy of spiritualism of the soul-force or love-force, which to him marked the Indian culture from the Western.

Gandhi considered Swaraj to be based on the freedom of every individual to be the master of the means of his livelihood. He was opposed to centralization of the means of production either under the sway of socialism or communism. He reinforced the strength of voluntarism in the economic aspect of national life by decentralization of political authority to the gram panchayats (village councils) , which were to be completely independent of provincial or central government. He intended to build the later on the foundation of the former. Gandhi's concept of Trusteeship would apparently make the capitalists hold the labourers in ransom. Vinoba Bhave has interpreted trusteeship to mean Vishwastabhava, i.e., a feeling of mutual confidence, not only between the propertied and the labouring classes, but also between the rulers and the ruled so that the rulers with all the democratic checks and balances and the governments' accountability to the people would discharge authority with a sense of responsibility towards the ruled. Gandhi wanted each agriculturist to own and operate the spinning wheel (Charkha) to supplement his meagre income from land and keep him fully employed. Not only that, he desired different village crafts and industries to process the produce of the village from agriculture, cattle or other sources. He was met oppose the use of Machines as such, but he desired the machine to serve the man rather then the man being servile to the machine. Voluntarism, the foregoing discussion, was at the core of Gandhi's thinking, on the reconstruction of India's economic and political organisation.

Another trend in Indian political economy before independence worked in the direction opposite to the trend of voluntarism. The exploitative nature of the economic implications of the British imperial policies in the country resulted in the insistent demands from the Indian Nationalist Movement for alternative protectionist (in the industry sphere) and welfarist in the agricultural and rural spheres and State developmental policies. In education, larger government grants - in - aid to the schools and colleges were asked for sizable State investment in technical education. In agriculture, lowering of land revenue assessment level, if not, stabilizing it at the prevailing rates, was pressed. Greater financial, technical and material help was expected from the government to raise productivity of land. Increased medical and health aid was demanded to stem the high death rate and to improve the standard of health of mothers, infants, and other people. These instances of demands for State intervention and consequent construction of laissez faire or voluntarism in development can be multiplied. At the time of country's independence, therefore, both these trends of voluntarism and State interventionism in development subsisted side by side in the economic and social spheres of national life.

Factors Promoting the Growth of Voluntary Agencies in India

Specific environmental factors that have promoted the growth of voluntary agencies in India can be summed up under your headings: Hindu traditions,

Gandhi's influence, Ideology of Left, and Perception of System Failure. A number of other facilitating factors has been operational as well.

Hindu tradition

Voluntary actions are performed of one's free will, impulse, or choice; they are not constrained, prompted, or suggested by another. These actions spring from a variety of sources; some lie within the individual, others in shared experiences.

In the western societies, voluntary action finds its expression typically in an organization, a formal and relatively easy to measure manifestation of collective action. Voluntary action in India, however, does not always take the form of an organization, many of our charitable customs are individual in their essence, for instance, the custom of ramroti to feed the Sadhu's is non-organized type of voluntary activity. While highlighting the non-organized nature of the activity, these customs also point to the caste-based character of philanthropy in India.

Charity flourishes in India. It is limited, however, by the Hindu world-view in three ways. First, it would seem that charitable acts are usually performed with the intention of discharging ones obligation towards a departed kinsman or to fulfil promises made to a deity. For e.g., a man may donate money to an orphanage so that his deceased father's soul may rest in peace or because he has taken a vow to do so.

Second, charity is confined to the religious space; it does not readily expand into the wider, secular space. For instance, going on pilgrimage is important for a traditional Hindu. Therefore, temples and shrines have to be kept in good repair and basic amenities provided to the devotees. It is customary for pious, wealthy families to build dormitories for the pilgrims and provide for their upkeep. Households with more modest means also offer their mite towards the same. Anecdotal evidence indicates that this type of activity dominates philanthropy in India.

Third, when not occurring as a religious ritual or in the religious space, charity takes place in the confines of kinship groups, since Indian cultural norms impose an obligation to tend to the needs of fellow-members of primordial groups such as the extended family and the caste.

Given the limited role of religion in inducing organized voluntary action, the source of inspiration for proliferating voluntary agencies must be sought elsewhere. Gandhi's influence, the ideology of the Left, and the perception of system failure suggests themselves as alternative sources.

Gandhi's Influence

The concept embodying the precedence of society over State was strongly reinforced by Gandhi's distrust of formal power, and was an integral component of both his vision of future India and the path he recommended for realizing that vision.

Second, Gandhi believed that with Independence, the Congress party had outlived its utility as a propaganda vehicle and a parliamentary machine'. He urged Congress members to leave politics for constructive work which he considered more important. Constructive work does not separate issues of material well being from those of spiritual well-being; it stresses social and cultural regeneration of the individuals as well as of society as a whole. Gandhiji exhorted Congress members to abjure power and dedicate themselves to pure selfless service to others. He suggested that constructive workers should not aim at entering Parliament, but should work to keep Parliament under check by educating and guiding voters . He called on the Congress party to turn itself into a Lok Sevak Sangh.

His call to the Congress party to disband as a political party went largely unheeded, but scores of his followers pledged themselves to a life of constructive work and austerity. Although very few combined the missionary zeal with a complete indifference to their own well being as did Vinoba Bhave, several hundred men and women founded groups and organizations to serve the poor while adopting spartan life-styles themselves.

Ideology of the Left

The second important source of voluntarism lies in the ideology of the Left. Even before Independence, the growing hegemony of the centrist faction in the Congress party had brought into open the ideological splits within the party. Consequently, some leftist factions in the Congress party established themselves as independent political parties and began mobilizing and organizing people. Later on, party workers, especially of the Communist party, became disenchanted with the party since it had resolved to function within the framework of parliamentary democracy and competitive electoral politics, attenuating its goals, strategies and tactics. Frustrated as the party workers were by these constraints, they were utterly disillusioned when they found that the party had failed to perform even this circumscribed role effectively. Consequently, many party workers severed their party connections and began to work independently in voluntary organizations

System Failure

In the 1960's and 1970's, a mood of disenchantment set in. Both the officials planning system and the market economy had failed to make a significant development in India's problems of poverty and inequality. This gave rise to scepticism about the ability of the institutional structures of democracy-legislatures, parties, unions, panchayats- to address the problems and needs of the poor. Formulated policies were not suitable enough to reduce poverty and inequality, and when they were, they were not implemented effectively. The government-sponsored model of development was seen as having failed to deliver benefits to the poor, and the formal political establishment had lost its legitimacy.

More specifically, the Congress party, a dominant, long-standing institution epitomizing democratic processes and norms, was crumbling. The deterioration of the Congress party disturbed the precarious balance that the Congress Party System' had maintained by accommodating diverse claims on the limited resources of the State. This decline in intermediate structures created a void in the process of political mediation between State and society. To fill the vacuum, a new class of social mediators' arose which led local protests and civil rights movements. Thousands of urban, well-educated young men and women, averse to electoral politics, swelled the ranks of activists founding or joining voluntary organizations.

Development of voluntary action from 19th Century onwards

An attempt has been made to divide voluntary action from 19th century into seven phases.

Phase One : The first half of the 19th century was marked by the initiation of the social reform movements. The charter act of (1813) finally rushed all restriction on missionary activities in India, which led to the expansion of missionary work in a big way. The formation of Atmiya Sabha in (1815) by Raja Ram Mohan Roy was one example, which later allied with Christian Unitarians and started the Unitarian Committee in (1821). The Spirit evoked by reform movements of devoting ones life to do something for the rights of the deprived and marginalised inspired a lot of peoples in this phase.

Phase Two : The Landmark of the second half of 19th century was the failure of "First war of Independence" in (1807), and the implications on the socio-political milieu. This also was the time of consolidation of British Colonial rule over the political and economic life of Indian Society. These trends also consolidated themselves in institutionised movements like Bramho Samaj. Ramkrishna Mission, Indian National Social Reform Association, (1897). The spread of such literary associations contributed to the development of an influential vernacular press and beginning of alternative education on the other. In a nutshell it is stated that during the second phase, (1850-1900) self help energized as the primary focus of socio-political movements and influenced the future course of voluntary action.

Phase Three: The major factor in the history of phase three has been the successful attempt of channelising the voluntary spirit for political action and mass mobilization for struggle of independence. The second factor was Gandhiji's initiation of "Constructive work activities" from (1920)-1928). This later became part of metro national movement and contributed to a number of Khadi and village industries.

Phase Four : This started after independence. The first 20 years of independence till mid sixties may be termed as nation building. Reform based voluntary actions and constructive work energized in the government responsibilities and tasks of nation building. The work of missionaries spread more in this phase with new institutions of education and health being set up in different parts of the country45.

Phase Five : This is seen between mid sixties to early seventies. This was the period when the development model followed by the government was critical and evidence about the failure of that approach to development become evident. By the late 1960's India was caught up in a dual crisis of economic stagnation and political instability. It was at this stage that alternative and integrated rural development began to be experimented by the initiatives taken by a new generation of people in 1968-69. This was also a period when Naxalite movement surfaced. The new professionally trained youngsters also began to enter voluntary development organisations.

Phase Six : This was the period when circumstances forced a number of people o reflect upon their experiences, to look back critically and the emerging trends of :ountry's political process. The process of politicisation of the past independence generation, which began during 1967-69 period was almost shattered by 1976. This fall out from the political process contributed to the growth in voluntary action, both in terms of Quality and Quantity. This was a period when ideas about conscientization and peoples participation began to emerge.

Phase Seven : It was the period of eighties, which witnessed a growth in voluntary action at other levels. A period, which gave rise to professionally trained social workers from different academic institutions joining the sector of voluntary action.

The voluntary action in different parts of India was ruled in a specific socio-political context and was inspired by the emergence and continuity of social reforms, social change and political movements in different parts of the Country. All along this period voluntary action had also matured and began to emerge in extension and conscientization and organization of reorganised people. What started as social work with a focus on charity has veered towards developmental work and community mobilization. There has been a proliferation of voluntary organisations across the board, each with its specific perspectives, priorities and strategies.

History of Voluntary Service in Maharashtra and Marathwada

In Maharashtra social, political, cultural and educational reforms were brought about by many political activists, philanthropist political activists and social reformers. They all worked selflessly, individually and organised groups. The works some of these leader is highlighted in this chapter, they are Dr. Babasaheb Ambedkar, Mahatma Gandhi, Agarkar, Mahatma Phule, Karve, Tilak, Pandita Ramabai, Shahu Maharaj, Sayaji Gaikwad, Deshmukh, Bhaurao Patil, Ranade. They all worked for the common man, the rural people, women education, upliftment of the untouchables and widow remarriage.

Pandita Ramabai (1858-1922)

P. Ramabai was looked upon as a symbol of love, humanity and philanthrophy. Sarojini Naidu said, "Pandita Ramabai was the first Christian saint of the Hindu series." She did the valuable work in Maharashtra's women education.

Vithal Ramji Shinde (1873-1944)

V.R. Shinde was the ideal of knowledge, humanity, patience, social work and living for social service. He worked for the upliftment of untouchable and for the education for women.

Maharaja Sayaji Rao Gaikwad (1863-(1939))

He was the King of Badoda State, though not the son of the soil. he worked for the Marathi people selflessly. He importantly worked for the betterment and development of the farmers in Maharashtra. He established the land survey settlement department. He also started girls education and industrial training.

Rajshree Shahu Maharaj (1847-1922)

During the difficult period of British reign after the unsuccessful rebel of (1857), Shahu Maharaj carried on his social work in Maharashtra. Due to his social reforms, he was called the King of people. Till his death he fought for the removal of social discrimination in the society. In 1857 he set up the ration foodgrains shop for people employment, co-operative act, lands for the cattle (gairan) and compulsory and free primary education law.

Justice Mahadeo Govind Ranade (1842-1901)

Justice Ranade was highly empowered judge, educated, historian, social reformer, politician and economist. He was one to the leading founder of National council, social conference and industrial council.

Gopal Ganesh Agarkar (1856-1895)

Shri Agarkar is one of the important social reformer in Maharashtra. In 1884 he established education society and in 1885 he established Fergussan college in Pune. He always fought against child marriage, castism and tonsuring, Keshwapan the head of widows. He also promoted marriage of choice for happy family life in the society.

Bal Gangadhar Tilak (1856-(1920))

Tilak and Agarkar both worked for people awareness in Maharashtra. Tilak was a National leader. He worked for Nationalism in India, untouchability, National education program in India. He also contributed for industrial business training in Maharashtra.

Gopal Krishna Gokhale (1866-1915)

Shri Gokhale major contribution was in the field of politics. He promoted overall development of society. His writings were helpful for people's awareness and their education. During his tenure as legislative member he tried for making education compulsory and free. Maharashtra Government established separate department for education with his inspiration. In 1905 he set up Servant's of India Society. Bharat Sevak Samaj in Pune . The objectives of this society were :

- To create patriotism.
- To practice service and sacrifice.
- Co-operation in all religion and castes.
- To organise political and educational movement.
- To work for industrial and scientific education; and
- To work for betterment of backward people.

Maharshi Dhondo Keshav Karve (1858-1962)

His concept was that there is no upliftment without cooperation. He established commission for business enterprises. He worked hard for women education, remarriage of widows and building up their confidence. He established orphanage for girls, commission for widow remarriage and selfless service schools. Social service is God' was the concept of this school. To educate one woman is to educate family was the concept used by Karve. He established Women Educational Centre at Hinge, Women's University and women schools.

Mahatma Gandhi (1869-1948)

The concept of his work was to teach moral values to students. His life was based on discipline, and justified principle. He was devoted to God and religion. He accepted that every life is a part of God and believed that service to nation will bring about salvation. His reign began after (1920). His three important principles were Satya]truth, Ahimsa]non-violence and Prem] love. He initiated the service for untouchables and established Wardha education program for India. The objectives of this program were:

- Compulsory primary education for all.
- Mother-tongue should be the medium of education; and
- Education should be imparted with profession.

He also worked for adult education. His teaching were valuable not only for India but for the whole world. T,herefor he is called the Father of Nation'.

Educational Reformers

Dr. Babasaheb Ambedkar (1891-1956)

The progress of any society depends upon the intellectual and able youths was the theme of Ambedkar. He worked for removal of caste system. He fought for human rights and prosperity of dalits. He was an enthusiastic leader. Ambedkar was well known lawyer, constitutionalist, politician and a champion for the dalits. He is the sculptor of Indian Constitution. He fought for untouchable, freedom movement and to bring about unity in the society. He established Schedule Caste Federation, Republican Party, hostel and schools to organise youths.

Karmaveer Bhaurao Patil (1887-1959)

Shri Patil worked for Phule's Satyashodhak Samaj. His concept was of self-dependent education. He proposed the thought of hard work. He established educational centre, training colleges and engineering colleges. His principles were of humanism, self-dependence, freedom, justice, service and sacrifice.

Punjabrao Deshmukh (1898-1965)

Shri Deshmukh worked for farmer's society. He reformed Krushak Samaj and established farmer's organisation. The objectives of this samaj were;

- To study the social, economic and cultural problems of farmers.
- To educate farmers in model farming by the experts.
- To form the national policies for the well-being of farming and farmers.

- To organise conferences, discussions, seminars, exhibitions on farming and to organise the farmers.

Shri Deshmukh organised National Krishi Co-operative for Trade, Bharat Krushak Sahakari Pat Pedhi, Yuvak Krushak Samaj, All India Honey-Bee Keeping and established Shri Shivaji Vidyalaya, colleges in various districts, Kasturba Kanya School for girls and Earn and Learn scheme (Kamwa and Shika Yojana).

Dr. Bapuji Salunkhe

Dr. Salunke worked for educational, cultural promotion and development of Maharashtra. He formed Rayat Shikshan Sanstha and Swami Vivekanand Shikshan Sanstha. He brought education to the common people.

Mahatma Phule (1827-1890)

Mahatma Phule was well-known social reformer of Maharashtra in the nineteenth century. He worked ceaselessly for education of the women and dalits, for upliftment of the underpriveleged and the downtrodden, and for reform of the Indian social structure. He was revolutionary in his thinking and is a constant source of inspiration for the new generation of intellectuals.

The rise of voluntary activities in Marathwada

Marathwada is a region of Maharashtra state. It comprises of eight districts, viz; Usmanabad, Latur, Beed, Nanded, Parbhani, Hingoli, Jalna and Aurangabad. Marathwada has a legacy of ancient history and culture. The saints, poets and social workers have great share in bringing about social and religious integration in Marathwada. Their teachings were great inspiration for the society.

The voluntary movement in Marathwada began with the struggle of Marathwada liberation and contribution made by Swami Ramanand Teerth in particular with the contribution and assistance of Govindbhai Shroff, Borikar, Charthankar, Deewan and many more.

Marathwada was under the rule of Hyderabad's Nizam for many years. Though India attained freedom, the Nizam of Hyderabad was unwilling to bring his State into India. This made the people of Marathwada to begin the struggle for liberation of Marathwada. This liberation struggle was led by Swami Ramanand Teerth. At last the Government of India with it's Police Action movement in 1948 helped the Union of Hyderabad into Indian State. The Nizam Meer Kamruddin established a State in Hyderabad in (1724) which had Marathwada, Telangana and Karnataka regions. He kept his State away from the Indian union.

Swami Dayanand Saraswati established Arya Samaj in 1875. Swamiji was a religious and social reformer. By 1930 every district had the organisation of this Samaj. The main functions of this organisation were social and religious reform. It also had an important role in the Marathwada Liberation struggle. The samaj tried to bring about humanity in the people with their social and religious reforms. This unity gave boost to the freedom struggle. This samaj brought about awareness in the Hindu community which was created through the newspapers. Swamiji also united the women force. With the efforts of Arya Samaj the spirit of patriotism was strengthened. The samaj strived for the upliftment of the dalits. It worked for all the religions and created unity and harmony among them. It was Arya Samaj which proposed the intercaste marriage and remarriage of widows. Solapur being convenient for educational and economic centre, the students of Marathwada went to Dayanand College for education there. It was this college which inspired the youth for the freedom struggle against Nizam. The youths were involved in Flag Satyagraha Movement and Vande Mataram Movement. The youths were inspired for freedom first and then education. Every district of Marathwada had the centre of Satyagraha.

Shri V.H. Desai, a veteran freedom fighter and an eminent journalist describes that it is only the liberation of Hyderabad that the Indian independence was complete in the real sense and this freedom struggle of Hyderabad or Liberation struggle of Marathwada was led by the dynamic leadership of Swami Ramanand Teerth. Swamiji harnessed all his energies for the most difficult task of reconstruction of the state. Reconstruction in all walks of life - social, political and economic. Swamiji was keen to involve the State Congress in the constructive programme of reconstruction. He had deep faith in equality and social justice. He had unanswering faith in socialism. He believed that the Indian polity and society should be reconstructed on the foundations of equity, social justice and secularism. He was the sole leader of State Congress from 1939 to 1948. Swamiji desired that this national political organisation should cast the people's life into new moulds so that true democracy could be evolved. He deeply involved himself in constructive activities and programmes. He formed the Maharashtra Council in 1937, which held its inaugral meeting at Partur taluka (Jalna district) . He looked upon education as a vital means of social regeneration and national reconstruction. He founded the People's College at Nanded in 1950, Medical College at Aurangabad, Yogeshwari Shikshan Sanstha at Ambejogai and also strived for the foundation of Dr. Babasaheb Ambedkar Marathwada University. He made four years of education free for the college going girls. He built hostels for them. In 1950 he started a english magazine Vision'. After 1950, swamiji did a lot of constructive work in Marathwada. He was greatly assisted by Govindbhai Shroff, Shri Paranjape, Shri Borikar, Shri Pedgaonkar, Shri Charthankar, Shri Chapalgaonkar, Janardhan Desai, Captain Joshi, Shri Vithal Joshi and many more.

Swamiji had following movements to his credit - Vande Mataram Satyagraha, State Congress Satyagraha, Individual Satyagraha, Struggle of (1942), leadership of Hyderabad Freedom Struggle, (1947) Movement against Law, Flag Satyagraha, Jungle (Forest) Satyagraha, Non Co-operation Movement, Bhoodan Movement etc.

The voluntary action in Marathwada was primarily relief work, education, development activities, agronomy, watershed projects and educating and promoting advanced farming skills. The poor economy coupled with drought prone poor yield land made development process an uphill task. The initiative of voluntary action was taken by individuals with the attitude of service filled with the spirit of philanthrophy. Like minded individuals formed trusts and worked for the upliftment of the underprivileged and development of rural and urban area of contact.

In Marathwada considerable work was done by Rotary club and Lions club. Rotary club in Marathwada was first established in Aurangabad District in 1952. They worked in the fields of medical assistance, education and relief activities. There is a unending effort made by the Parsi communities in the development of Marathwada especially in Jalna District. They established schools and colleges for promoting education. They generously donated money and land for the poor farmers in the region. The popular project of Ghanewadi which was initiated by the British army to built catchment area due to the water problem and left incomplete was completed by Bezonji himself. This project now serves as the main source of water in the region.

The Christian missionaries came to Marathwada initially for development and relief work and educational facilities. They set up schools for boys and girls separately. They established hospitals and worked in the urban and rural areas doing both curative and preventive medicines. They realised that unless rural economy is increased, there will be negligible development and progress in the area. The War-on- Want was set up in 1957-1958, this organisation worked for drilling bore wells, watershed projects, building roads and constructing bridges all which increased the productivity of the land and increased the economy and improved the transport and communication. In 1961-1962 John Macleod and Dr. Wigglesworth worked tirelessly to increase the water level in Marathwada by digging wells, bore wells, constructing cathment areas and water bunds, the philosophy of watershed was thus born. They devised new techniques for boring wells, which was the pioneer model for all boring apparatus nationwide.

1961 onwards Shri Barwale worked with the War-on-Want to develop the villages of Mandava, Bhokardan etc by extending credit schemes to the farmers through Bhoo Vikas Bank, digging wells and supplying super quality high yield seeds for farmers. Then Mahyco a private sector was formed, which developed and experimented with hybrid seeds. The farmers were also educated in scientific, developed and modern methods of agriculture.

In 1962 Shri Paranjape and Shri Nade worked for gramin vikas in Murad (Osmanabad District) . In 1967-1968 Marathwada Sheti Sahaya Mandal was formed with the objective of helping the marginal farmers and landless people. The amalgamation of Mandal and War- on - Want in 1971-1972 saw the development of integrated agro services programs. The objectives of this program were to extend credit for farmers and to educate them. This utilisation of credit was through MSSM and War - on -Want. The collaborated efforts of Mahyco working in agronomy, War-on- Want in water harvesting and watershed and MSSM in the field with farmers, educating them was the first of the cooperative efforts in rural development of Marathwada.

In 1970-1972 Marathwada experienced worst famine in the history. Many individuals, trusts, Christian Missionaries and Churches,Church Auxillary and Service Action (CASA), Chruch Integrated Service (CIS), MSSM, Mahyco, etc worked in the villages of Marathwada on a war footing. Feeding program for the under five year children and mothers, medical help, digging wells, supply of potable water, credit to farmers, distribution of cattle and fodder were some of the projects undertaken. The Missionaries worked for relief activities and *Garibi Hatao* Programs. The relief activities were multifold. It was economic support on one hand and education and health on the other. Macleod strived for water availability in all the villages and he was also instrumental in building roads into the villages so that there was proper communication and trade between the villages and urban areas.

Narayan Sheshadri worked in the villages of Bethel, Revgaon of Jalna District developing the area, building check dams, establishing tannaries, cottage industries and fruit orchads. He also built roads, schools, and hostels along with veterinary hospital, poultry farms, bee keeping and animal husbandry. Government established Dharam shalas in 1970 for the poor and needy. All the programs in the region floated were development oriented.

Typology of Voluntary Agencies

In India, societies, associations, organisations, trusts or companies registered under the Societies Registration Act, 1860; the Indian Trusts Act, 1882; the Charitable & Religious Trusts Act, (1920) or as a charitable company under Section 25 of the Companies Act, 1956 are considered as VOs / NGOs. In addition, there are informal groups working at grassroots level without being registered under any legislation but may also be considered as part of voluntary sector. NGOs may be working in the field of welfare of disabled; development of other disadvantaged sections like SCs/STs, children and women in education; environment; human rights; and on issues like resettlement and rehabilitation of ousters by big projects, right to information and so on. VOs may take up issues concerning a particular village or a community to the global issues like impact of WTO or global warming. The range of associations or societies may vary from a resident welfare association to an advocacy organisation. The substantive areas of work of VOs have changed considerably over time. Recently, in an exercise to map the diversity of voluntary sector in India, has provided the following typology of voluntary associations.

Traditional Associations

Such associations exists around a social unit either defined by a tribe, ethnicity or caste. Associations of this variety undertake a wide range of functions in the lives of those citizens. Besides mediating inter-family relations, such associations develop elaborate systems, norms and procedures for governing the use and protection of natural resources. Several important struggles to protect and advance customary rights of tribals over natural resources in different parts of the country have been led by such associations.

Religious Associations

Over the centuries, new sects and religions were born and incorporated into Indian life. Buddhism, Jainism, Sikhism and many other variations challenged Hinduism, Judaism, Christianity and Islam for reform and renewal. Charity, help to the needy, service to the poor and daan (giving) have been uniformly recommended by all these religions and sects in India. Education, health care, drinking water, afforestation, social welfare, etc. are numerous arenas of human action where fairly organized forms of civil society activity are carried out by religiously inspired VOs.

Social Movements

In the contemporary Indian context, a number of social movements spearheaded by social movement organisations (SMOs) have emerged as major manifestations of civil society. These movements are of several types:

- Focusing on the interest and aspirations of particular groups such as : SCs, STs and women.
- Protests against social evils like: liqour, dowry, inheritance rights etc.
- Protests against displacement due to big development projects.
- Campaigns against environmental degradation, corruption and for rights to information, education and livelihood.

NGOs through these social movements try to reform society, institutions and governance and act as harbingers of social change.

Membership Associations

Membership organisations may be representational - set up to represent the opinions and interest of a particular category of citizens e.g. unions of rural labour, farm workers, women workers, consumer associations etc., professional - formed around a particular occupation or profession e.g. association of lawyers, teachers, engineers, managers, journalists etc., social-cultural - organised around a social or cultural purposes e.g. Nehru Yuvak Kendras, clubs for sports, Natak Mandalis etc. and self-help groups - a growing category of membership organisations e.g. savings and credit groups etc.

Intermediary Associations

These associations function between individual citizens and macro State institutions like the bureaucracy, judiciary and police etc. These could be of several types, e.g. service delivery, mobilizational, support, philanthropic, advocacy and network.

Role of Voluntary Agencies

NGOs imbued with the Gandhian philosophy were playing significant role mainly in social welfare activities and now, the range of spheres covered by voluntary sector has expanded considerably covering almost all development related activities. Some of the reasons put forward for increase in the number and activities of NGOs are the decline of socialism and an increased national and international funding for voluntary sector. In a number of developmental activities, NGOs are working as supplements or complements to the governmental efforts.

The activities and role of NGO's can be considered under three heads: developmental, political and catalic. Developmental activities aim at the poor directly. They comprise the delivery of a wide range of services and take various forms. Voluntary agencies may (a) actually deliver the benefits; (b) they may act as a bridge, providing information about relevant government schemes to the poor; or (c) they may help target groups meet procedural requirements so that the poor may reach up and pull down to them the benefits of development'.

Voluntary agencies perform the role of a catalyst. A catalytic activity aims to influence public in a way which the NGOs expects will initiate action. This activity may launch a public campaign to dessiminate information.

Political activities, on the other hand, are directed at a governmental authority. It aims to persuade the authority to take action to create such conditions as would improve the lot of the poor. When a voluntary agency convinces the municipal health department, a public authority, to launch a scheme to inoculate children in city slums, it is engaging in political action. Political activities are important because they attempt to bring about changes at the policy level. If implemented effectively, the policy-level action can have far-reaching consequences since it encompasses all that fall in the defined category.

Developmental, catalytic and political activities are not mutually exclusive, and most voluntary agencies pursue some combination of all three, often in conjunction with each other or as a series of successive steps. However, both voluntary agencies themselves and the researchers have shied away from focusing exclusively on political activities.

References

Michael Banton, "Anthropolitical Aspects, Voluntary Associations" in David Sills (ed) International Encyclopaedia of Social Science, Vol 16, New York. The Macmillan Company and the free press, London, Collier Macmillan 1968. P. 357.

David L. Sills op.cit, p.362-363.

Marilyn Gittel, Limits to Citizen Participation', Sage Library of Social Sciences, 109, P. 22-23.

Constance Smith and Ann Freedman, Voluntary Associations perspective on the Literature, Cambridge (mass Harvard University Press, 1972.

A.F.C. Bourdillon (ed) Voluntary Social Services : Their Place in Modern State. London, Methuen, 1945. P.2.

William Beveridge, Voluntary Action in a changing world, London, Bedford Square Press, National Council of Social Services, 1979. p. 1-5.

Norman Johnson, Voluntary Social Services, Oxford, Basil Blackwell and Martin, Robertson, 1981, p. 14.

Seventh Plan, Government of India, para 2, p. 115.

Constance Smith and Ann Freedman, op.cit, p. VIII.

D.L. Sheth and Harsh Sethi, The NGO Sector in India : Historical Context and Current Discourse,' Voluntas 2, 1991. p. 51.

Rohini Patel, Voluntary Organisations in India : Motivations and Roles', in Social Change Through Voluntary Action (ed) Chap. 3. p. 41.

Richard Taylor, The Encyclopaedia of Philosophy, (Paul Edwards, Editor-in-Chief Vol. 8 New York, The Macmillan Company and the Free Press London, Collier Macmillan 1967 p. 270.

M.A. Muttalib, Voluntarism and Development - Theoretical Perspectives : The Indian Journal of Public Administration Quarterly Journal : Special Number on Voluntary Organisations and Development, Their Role and Functions, July-September 1987, Vol XXXIII, No. 3.

A.F.C. Bourdillon, op.cit, p. 7.

G.D.H. Cole, Voluntary Social Services', (Edited by A.F.C. Bourdillon London. Methuen. 1945. p.11-12.

William Beveridge, op.cit, p. 100.

G.D.H. Cole, op.cit, p. 16.

Ibid, P. 24-28 : Compare British Experience in this Respect as a Specific Example.

Ibid, p. 41.

H. Hansmann, The Role of Non Profit Enterprise', *Yale Law Journal* 89, 1979-80, p. 835.

F. Fama and M. Jensen, Seperation of Ownership and Control', *Journal of Law and Economics* 26. 1994. p. 301.

Peter B. Clark and James Q. Wilson, Incentive Systems : A Theory of Organisations.' Administrative Science Quarterly 6, (1961), p. 129-66.

N.R. Inamdar : Role of Voluntarism in Development, *The Indian Journal of Public Administration Quarterly Journal* 6. (1961). p. 129.

A.S. Altekar, State and Government in Ancient India, Delhi, Motilal Banarasidas, (3rd Edn 1958. (reprint 1972 p. 384.

Seventh Five Year Plan 1985-90, Vol II, Delhi, Government of India, Planning Commission, 1985, P. 68-70.

N.R. Inamdar : The Pattern of Public Administration in Kautilya's Arthashastra' Oriental Thought, Nashik, 1956.

Altekar, op.cit, p. 355.

M.K. Gandhi, Hind Swaraj or Indian Home Rule, Ahmedabad, Navjivan Publishing House, (revised edn (1939). Reprint 1982 (first edn. 1938 p. 104.

Young India, November 15, 1928, p. 381.

Vinoba Bhave, Introduction in K.G. Mashurwala, Gandhi and Marx', Ahmedabad Navjivan Publishing House, 1951. p. 24-26.

J.B. Kriplani, Gandhi - His Life and Thought, New Delhi, Publications Divisions, Ministry of Information and Broadcasting, Government of India (second reprint 1975 p. 377-378.

Jathar and Beri, Indian Economics, Vols 1 and 2 (Oxford University Press D.R. Gadgil Industrial Evolution in India (O.U.P. N.R. Inamdar] (ed) Political thought and Leadership of Lokmanya Tilak, New Delhi, Concept 1985; Vide Inamdar's Introduction and Bureaucracy and Political Ideology of Lokmanya Tilak'.

The Oxford English Dictionary] Oxford : Clarendon, 1989 p. 754.

K.M. Sen, Hinduism'] Baltimore : Penguin Books (1961) p. 23-24.

Robert L. Hardgrave, Jr. and Stanley. A. Kochank, India : Government Politics in a developing Nation] San Diego : Har Court Brace, Jovanovich Publishers, 1986 p. 52.

Sheth and Sethi : The NGO Sector', No. 4.

Bharat Jhunjhunwala, Voluntary work as Contervailing Power', Economic and Political Weekly 21, 1986, p. 599.

Sheth and Sethi : The NGO Sector', No. 4, p. 52.

Harsh Sethi, Groups in a New Politics of Transformation', No. 7.

Sheth and Sethi : The NGO Sector', No. 4, p. 50.

Leslie J. Calman, Toward Empowerment : Women and Movement Politics in India] Boulder, Co : West View, 1992 p. 21.

Amrita Basu, Two faces of Protest : Contrasting Modes of Women's Activism in India] Berkley University of California Press, 1992 , p. 16.

Subrata Kumar Mitra, Crisis and Resilience in Indian Democracy.' *International Social Sciences Journal* 43, 1991. p. 565.

It must not be overlooked that they were unable to find employment in the Public or Private Sector.

Alka Srivastava : Non-Governmental Organizations and Rural Development : Social Action; A Quarterly Review of Social Trends; A Social Action Trust Publication, Vol 49. January-March 1999. p. 29-30.

Alka Srivastava, op.cit, P. 31.

Anna Saheb Garud and B.B. Sawant, History of Social Reforms in Maharashtra from 1818-(1950), Kailash Publications, 1995. p. 98-102.

Anna Saheb Garud, op.cit, p. 144.

A.M. Kathare and Dr. N.N. Nagrale, History of Marathwada, Beginning to (1960) : Kalpana Publications : 1999] marathi edn p. 56.

Ibid, p. 173.

Ibid, p. 185.

Ibid, p. 4-5.

V.H. Desai, The Visionary, Important Speeches and Writings of Swami Ramanand Teerth, Published : Swami Ramanand Teerth', Marathwada University, Nanded, 1996. p. XX.

Interview with Shri J.M. Gandhi, Director of Marathwada Sheti Sahayya Mandal, Aurangabad. He is a pioneer of voluntary service and watershed projects in Marathwada.

Report of the Steering Committe on Voluntary Sector for the Tenth Five Year Plan] 2002-2007 . Planning Commission Government of India, January 2002, Chap. 1, p. 3.] source - Internet

The term Catalyst' has been used in two different senses. Terry Fillibard] 1983 uses it to refer to the dispersibility of the organisation. Once a change is initiated, the organizer, who is almost an Outsider', withdraws from the group to avoid fostering dependancy. The agent is only a catalyst for change. Alternatively, the term refers to the mediating role voluntary agencies play between the state and the people, creating inter-linkages between the two by facilitating the two way flow of information, know-how and understanding. This is how Kishore Saint] 1974 defines the role of voluntary association. My usage of the term departs from both.] Rohini Patel, op.cit p.50

Rohini Patel, op.cit, p. 50-51.ji

CHAPTER-2

Voluntary Agencies and Rural Development

Introduction

This chapter includes the importance of rural development in India, stating the concept, meaning and various aspects of rural development. It focuses the role of NGOs in rural development with its types, characteristics and various activities in rural areas.

Conceptof Development

Development, as a term, signifies growth or change for the better in any aspect of a social process. Applied to an economy, it means growth in its different sectors of production and distribution, improving standards of education, living and civilization of the people, improving wages for the workers etc. Political development similarly means healthy growth of political institutions in society like democracy, public administration, growth with justice, equality among citizens, integration among communities and linguistic regions, and so on.

Social development signifies the development of social institutions which may facilitate smooth changes and outlook of people. It means improvement in social structures like norms of mutual conduct among members, values, cultural tastes, humanization and modernization. The most important agency of social development is educational institution spread in a society - both formal type and mass communication.

Hahn-Been Lee says : "Development is a process of acquiring a sustained growth of a system's capability to cope with new, continuous changes toward the achievement of progressive political, economic and social objectives."

Milton J. Esman rightly states : "Development is the national progress of organizing and carrying out prudently conceived and staffed programmes or projects as one would organise and carry out military or engineering operations."

E.X. Sutton says " "The goals and methods of development must mean a vision of rationalization... Rationalisation is a more consciously directed process, resting on comprehensive scientific analysis of what must be done and the deliberate training of men for new tasks and new ways of behaving."

Riggs Concept of Development

Riggs defined development as a process of increasing autonomy (discretion) of social systems, made possible, by rising level of diffraction.' Discretion', he observes, is the ability to choose among alternatives' while diffraction refers to the degree of differentiation and integration in a social system. Ecologically development is increasing ability to make and carry out collective decisions affecting the environment.

Riggs considered differentiation and integration as the two key elements in the process of development. Differentiation means existence of a situation in which every function has a corresponding specialised structure for its performance. Integration means a mechanism to tie together, to link up, to mesh, to coordinate the various kinds of specialised roles. The levels of differentiation and integration represents diffracted and prismatic conditions of development. If the society is highly differentiated and poorly integrated, it is prismatic. Diffraction leads to development, and the higher the level of differentiation and integration, the greater the level of development, and the lower their level, lesser the development. In the same way the level of malarrangement between differentiation and integration results in the different levels of prismatic conditions. Riggs explained the differentiation and integration correlations and resultant diffracted and prismatic conditions as shown in Fig. 1.

Riggs drew two lines of coordination and the changing levels of differentiation and integration. He pointed out a diagonal between the two representing the ideal level of integration required to handle the complexities involved in coordinating differentiated roles while providing sufficient autonomy for each role to be carried successfully and its distinctive functions. Thus, the line m, n...t represents a condition that is more diffracted than situations symbolized below the line a, b...g. On this scale the extremes represent mere hypothetical constructs, but the intermediate points come closer to characterise empirical realities.

Riggs also hypothetically presented imaginery societies which are becoming more and more differentiated without successfully mastering the problems posed by these changes through its integrative mechanisms. The shift from n to b or p to d in the graph represents this condition. The fact of having a variety of differentiated roles can led to greater confusion and chaos - unless the specialised roles are carefully coordinated with each other. There must be a mechanism to tie together different kinds of specialised roles. Integration, thus, becomes a highly essential part of the whole scheme. Riggs observes that "it is much easier to train people to perform the specialised roles of modern government than it is to integrate these roles, to link them up together." Development would be possible only when the roles are carefully integrated.

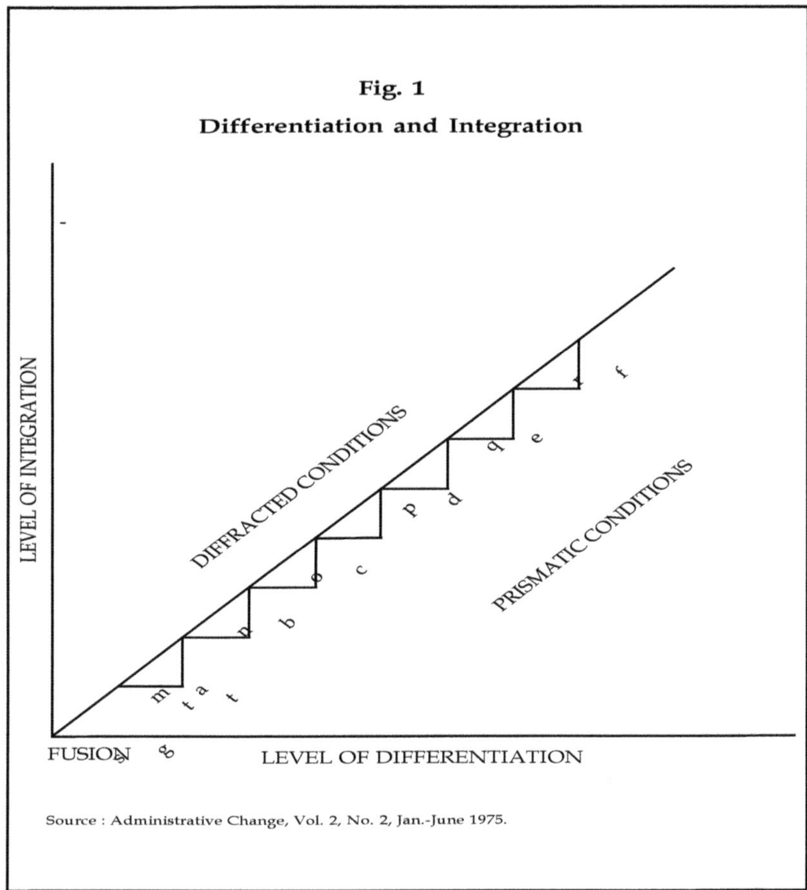

Fig. 1

Differentiation and Integration

Source : Administrative Change, Vol. 2, No. 2, Jan.-June 1975.

Riggs further explained his concept of development by analysing the factors affecting differentiation and integration. Chart 2 facilitates a better understanding of his concept of development.

The level of differentiation in any country depends upon the technological and non-technological factors. The more the development of technology, the higher the level of differentiation. The integration depends on the important factors: (i) Penetration, and (ii) Participation. Penetration is the ability of a government to make and carry out decisions throughout the country. Participation is the receptivity to law and the willingness to help carry out the laws and the policies which government has formulated. Participation, thus, has two important elements (1) willingness of the people to participate, and (2) ability of the people to participate. The more the willingness and ability to participate on the part of the people, the higher the level of participation in the governmental affairs. Thus the penetration and participation facilitate the integration of differentiated structures resulting in development[5].

Nature of Development

The most important component of development is its dynamism. Development must signify continuous change in society for the betterment of the people in their various phases of life. Development brings in its processes continuous change in the pattern of living of the people in different walks - spiritual, material, religous, social, economic and political. It is in the nature of development of influence the inter-relationships among different members of society, different groups or their combinations. Another important nature of development is that the changes which development brings in its wake take upon themselves the characteristic of unending cycle impact. One change causes another, the second multiplies into many others, and so on. This process of dynamic changes is an essential nature of development.

Prerequisites for Development

Since development is a continuous process the existence of certain prerequisites for development has to be ensured by :

A Stable Political System : If a society aspires for a planned development it must have the basic infrastructure laid down in an acceptable political order and a well defined constitutional system. Without such a base no society can move in the direction of nation-building and socio-economic progress. It must have a stable constitution to provide for mutual adjustment among its different social components, sections, classes and sectors.

Stable Administrative System : Unless a society has a well-regulated system of public administration it cannot undertake any programme of development for improvement in the environment of society. A system of public administration must have a sound base of rules and regulations, division of responsibilities, delegations of authority, division of work among different departments and an efficient system of co-ordination among different wings of the system like the executive, the judiciary and the legislature.

A Well-grounded Cultural System : Unless there is an environment of minimum cultural understanding, the heterogenous elements of society would always be clashing among themselves and instead of paving the way for the development of civilisation they would be tending to destroy or retard its development. An environment of integrity - social, political and cultural - is essential for economic development.

A System of Planning : It is necessary for a society to improvise the concept of planning; without a system of sound planning and the necessary machinery for the implementation of plans no society can achieve necessary developments or bring about modernization in its different walks of social life. Nation-building and socio-economic progress require a sophisticated system of planning in which there is well-regulated utilisation of the resources - material and man power[6].

Importance of Rural Development in India

It is a well-known fact that in ancient times, the rural people of India were organised into self-sufficient, hard-working, cooperative, happy village republics.

They were governed by panchayats democratically constituted by the community. The village system, however, was utterly destroyed during the British regime and the rural masses became helpless and exploited.

To give a dignified life to all her citizens living in rural and urban areas, independent India has pledged to establish a socialistic pattern of society' through planned development. Rural development, therefore, occupies priority in the agenda for national development in India for may reasons many.

India's 75 per cent population depends upon agriculture for their livelihood[7]. It is the single largest and important sector of our economy, which contributes about 40.9 per cent of the country's GNP. Agricultural production, due to its erratic nature, fluctuates more violently than industrial production. This fluctuation further leads to higher fluctuations in agricultural prices as well as agricultural incomes. The small and marginal farmers cannot bear the consequences of fluctuations in farm output, prices and adverse effects of free market. Only rural development can save them from their miserable conditions and can play an important role in increasing the purchasing power of the farmers. This increased purchasing power will demand for manufactured articles and therefore a big market for industries can be provided. In this way, rural development will also be in the interest of industrial development.

The politico-economic, and socio-cultural structural disabilities threaten the development of the rural poor. This has aggravated the problem of poverty, which has manifested itself in various forms, i.e., unemployment, inequality, malnutrition, growth of slums, and ignorance. The growing poverty, which undermines planned development, is much more severe in rural than in urban areas and has become the biggest bottleneck in national development. About 55 per cent of rural population in India is living below poverty line. The poverty of Indian masses, especially those dependent on agricultural, is proverbial. Weak, unorganised and unregulated landless agricultural labourers, marginal farmers, heavily indebted small farmers, artisans, and share-croppers are hungry and poor because of the existing structure of socio-economic and political power in rural areas. Due to poverty, illiteracy and hunger, man has been dehumanized; degeneration has set in; crime in the society has multiplied and vested interests become more and more stronger. Since rural development and poverty alleviation are two sides of the same economic problem, the former is essential to accelerate the pace of overall economic development.

Poverty arises from unemployment and the latter increases the former. Rural unemployment and under-employment are serious problems in India. They are on the increase due to big increase in population and the requirement of lesser labourers due to the latest technology. The existence of joint-family system and seasonal agricultural operations are mainly responsible for rural underemployment. The setting up of new industries, expansion and modernization of existing industries, small-scale industries and cottage industries in rural areas can create additional job opportunities to provide gainful employment to the unemployed and underemployed rural masses. Thus, to solve the chronic problem of unemployment and underemployment, rural development is the most important need of the hour.

Today, the advancement of material development and even the basic facilities and services in India, i.e., big hospitals, higher level schools and colleges, better transport services, civic and social amenities are concentrated in cities, where only about 24 per cent of the total population of India lives. As a result, a large number of rural people are migrating to the urban areas in search of better living and to earn livelihood. Sometimes, they do not get employment even in urban areas, which leads to semistarvation, beggary, dacoity, smuggling, terrorism and other social evils. In the process, many parts of the urban areas have been converted into slums, threatening the environment. The solution to many of these urban problems ultimately lies in the improvement of rural life. Thus, rural development involves urban development too.

Due to lack of resources and little or no bargaining power the owners of small, scattered and unorganised rural enterprises are exploited to the hilt vis-a-vis those to whom they sell their products and from whom they purchase raw materials etc. These small rural entrepreneurs can be protected from exploitation and their financial conditions can be improved only through rural development in the form of price support, credit facilities, insurance, etc.

Finally, rural development holds the key to nation-building. It is rural development, which can transform rural society from traditional isolation to integration with the mainstream of national life. It will evolve a new society with a built-in value system by raising the social, economic and cultural values of the rural people through basic infrastructural amenities, i.e., pure drinking water, nutritious meals, clothes, houses in clean environment, education, medical facilities, recreational facilities, electricity, metalled roads, avenues for cultural development, transport, communication etc., for their harmonious growth. The development of living and working conditions of the Scheduled Castes, Scheduled Tribes, backward classes, artisans, agricultural labourers, women as well as removal of the rising criminal incidences, plunder, untouchability, parochialism, provincialism, etc. will definitely bring harmony, peace, pleasure and contentment to their lives. Awareness among rural masses, their active cooperation and involvement in the development programmes will stop corrupt practices, black-marketing, smuggling, terrorism and other unwanted acts and thus contribute to the building of a strong modern India.

Concept of Rural Development

Rural Development has been in and out of the notice of political decision-makers many times, but again it is in and this time it is expected to be an integrated effort. Rural development is an elastic concept and everyone interprets it in his own ways, but the broad consensus is that more emphasis should be given to those rural development activities, which mainly concern the rural areas. These include agricultural growth, the putting up of an economic and social infrastructure, fair wages as also housing and house-sites for the landless, village planning, public health, education and functional literacy communication, etc. Thus, it focuses attention on two aspects :

- Economic development with a close interaction between other different sections and sectors, and
- Economic growth specifically directed to the rural poor. The thrust of attention in all the special programmes is towards the rural poor, not only in

> terms of providing incentives for development but also linking of economic activities into a well planned infrastructure. In this multi-pronged effort involving development and consideration of resources land, water and human, the objective is to raise the standard of living and the quality of life, particularly of the rural poor. Rural development, thus, encompasses both the spatial and functional integration of all relevant programmes bearing on increased agricultural production and also the reduction of unemployment, under-employment and provisions of gainful employment among the rural people.

According to a World Bank paper, rural development "is a strategy to improve the economic and social life of a specific group of people, the rural poor including small and marginal farmers, tenants and the landless." The term rural' means an area which is characterised by non-urban style of life, occupational structure, social organisation and settlement pattern. Development is defined in terms of technological or industrial development. But development of rural people means raising the standard of their living. It is the development of rural areas through the extension of irrigation facilities, improvements in the techniques of cultivation, expansion of electricity, constructions of school buildings, provision of education facilities, health care and roads, etc.

Uma Lata defined "rural development in terms of raising standard of living of rural people". One sociologist, J.H. Copp, has defined rural development as a "process through collective efforts, aimed at improving the well-being and self-realisation of people living outside the urbanised area. He further contends that the ultimate target of rural development is people and not infrastructure and according to him one of the objectives of rural development should be to widen people's range of choice".

According to a UN Report, "Rural development has come into international usage to connote the process by which the efforts of people themselves are united to those of governmental authorities to improve the economic, social and cultural conditions in the life of the nation and to relate them to contribute fully to national programmes".

Integrated rural development had been defined by Sharma and Malhotra as a systematic approach aiming at total development of the area and the people by bringing about the necessary institutional attitudinal changes and by delivering a package of services through extension methods to encompass not only the economic field, i.e., development of agriculture, rural industries, etc., but also the establishment of the required special infrastructure and services in the areas of health and nutrition, education and literacy, basic amenities, family planning etc., with an ultimate objective of improving quality of life in the rural areas.

In the words of Robert Chambers, "Rural development is a strategy to enable a specific group of people, poor rural women and men, to gain for themselves and their children more of what they want and need. It involves helping the poorest among those who seek a livelihood in the rural areas to demand and control more of the benefits of rural development. The group includes small scale farmers, tenants and the landless."

Thus, rural development is a multi-dimensional process which includes the development of socio-economic conditions of the people living in the rural areas, and ensures their participation in the process of development for complete utilisation of physical and human resources for better living conditions. It extends the benefits of development to the weaker and poorer sections of rural society. It also enhances both the capacity and capability of administrative and socio-economic development agencies and agricultural marketing units working in the rural areas.

Rural development Approaches

Area Development Approach : Under this approach emphasis is laid on the development of the underdeveloped regions. The area development approach presumes that the growth centres have an even geographical spread effect and that the benefits of development percolate to the lower levels over a period of time. Under this approach, a pinpointed area is taken for development. A backward area is identified for concentrated efforts, such as DPAP, TDP, CAD, Hill Area Development, etc., which comes under this approach. Integrated Area Development requires not only detailed action, cooperation and support from credit and service organisations in several disciplines/fields like irrigation, soil conservation and agricultural extension but also the building up of the basic infrastructural facilities in the fields of communications, irrigation, land development, processing, marketing, etc.

Spatial Planning Approach : The need for appropriately locating all the special programmes in their respective fields, the induction of production plans, the full-employment schemes and the supply of the basic needs of the rural population, all demand that the plan formulation and implementation strategy should be more rural oriented. In the Fifth Plan, multi-level planning was very much emphasised and it was argued that since more intimate, precise and detailed knowledge about physico-geographical, techno-economic, socio-political and organisational administrative conditions is available planning for activities which have strong local foci is more fruitfully undertaken at the district level,[17] therefore, under this approach the progress was too slow owing to a number of other factors affecting the national plan formulation.

Integrated Rural Development Approach : The Integrated Rural Development Approach has a unified field unlike the multipurpose approach. It is a plan of action. It comprises four types of activities, namely, increased production in agriculture and allied sectors such as animal husbandary, fishery, forestry and horticulture. It lays emphasis on village and cottage handicrafts and tiny industries; the tertiary sector which would cover artisans and the requirements of skilled workers in several rural activities and, finally, labour mobilisation which includes training in skills and organised employment for labour class. As T.N. Chaturvedi says, "It stands for the development of the rural society in all its facets - social, economic, institutional and administrative. The new approach stands for integrated' performance and accomplishment of all the objectives stipulated."

The Synergistic Approach : An earliest and still continuing approach to national development in India is the Synergistic Approach, which calls for the harnessing of modern science and technology for removal of country's economic and social backwardness. The Science and Technology Policy of the government reflects the firm faith of the political leaders and administrators of India to use scientific knowledge and technology in a big way to achieve then development goals including rural development.

Rural Developmental Strategies

Rural development strategy can be defined as a set of goals, operation processes, terminal objectives and structural arrangements designed to bring about change and development in the lives of the rural people. A typical rural development strategy is defined as one "that achieves desired increase in farm output at minimum costs, makes possible widespread improvements in the welfare of the rural population, contributes to the transformation of a predominant agrarian economy, and accelerates a broader process of socio modernisation." Rural development strategies, therefore, outline the processes that lead to a rise in the capacity of the rural people to control their lives and environment, accompained by wider distribution of benefits resulting from such control.

Typologies of Rural Development Strategies

Inayatullah classifies the rural development strategies in terms of the level of intervention with respective broad policy goals where productivity, solidarity and equality are considered as broad indicators. The models are :

Low intervention Productivity Model (LIP) ,

Medium Intervention Solidarity Model (MIS) , and

High Intervention Equality Model (HIE) .

Low Intervention Productivity Model

This model primarily seeks to raise productivity without necessarily bringing about significant changes in the social struture and land revenue system. The strategy of this model is to assist those who have the necessary capital, resources, skills and motivation to raise productivity. The model allows, to a great extent, the production priorities to be determined by the market forces also. The intended beneficiaries are the large farmer who, according to this model, can mobilise the necessary inputs for greater productivity.

Medium Intervention Solidarity Model

This model seeks to remove the bottlenecks through creation of new institutions, modernisation of the rural elite, and diffusion of organisational and human relations skills. It encourages co-operative activities by creating community-based credit, marketing and consumer co-operatives. Through these co-operatives, it attempts to mitigate exploitation in rural areas and to strengthen the economic power of the rural producers and consumers. The main beneficiaries of the model are those belonging to the rural middle class.

High Intervention Equality Model

The primary goal of this model is to narrow down and consequently eliminate the social, economic and cultural inequalities and exploitation of the poor by the rich classes. Rural development, according to this model, therefore, requires a two-fold struggle: control of the State appartus at the national level through appropriate political organisations by the rural poor to facilitate radical changes, and an organised struggle against the rich at local level through which the poor will enter the power structure and control resource bases and means of production.

An Historic View of Rural Development

In India 75 per cent of the population lives in rural areas and nearly 70 percent of the workforce depends upon agriculture and allied activities. The contribution of the rural sector to national income is substantial. Such an important sector will naturally draw the attention of the people and the government for its reconstruction. Several eminent persons have contributed their mite for rural development. Therefore, it is necessary to know about the past attempts which have given the present shape to rural development programmes. A historical account may also give a clear understanding of the dynamics of the programmes, the changes which have occurred, the direction that these changes have given. Further, the present evaluation and diagnosis of the weaknesses and also the suggestions for improvement will not prove useful if we fail to take into account the hopes and aspirations of the people affected by these programmes.

India has a long history of rural development efforts and experiments, which can be broadly discussed under three periods:

1. From 1858 to 1919 : The British Rule.
2. From (1920) to (1950) : Rural Development during National Movement.
3. From 1950 to 1976 (Since Independence).

First Period : 1858 to 1919

In 1858, the British Government took over the Governance of India from the East India Company, after the first attempt to gain Independence by the Indian in (1857). The British were interested in the governance only and they were not concerned with the socio-economic development of the people of India. The political thinkers of this period were of the view that the government which rules the least is the best. But it was the famine of 1899 which forced the British government to think about the people of India who were dying of hunger. The recurrence of famines again and again forced the state adminstrative machinery to seize the foodgrain stocks and make arrangements for the distribution the famine-affected people. All this was done without any apparent legal sanction, motivated as the action purely was by determination to control profiteering and alleviate rural suffering. As B.B. Mishra further says, "Rural development, thus, began as a humanitarian act, and the practice was not backed by any executive or legal sanction in the beginning. Legality and legitimacy were provided later. Rural development as a function of government initially began as a search for an alternative to Laissez-faire.

The famine Commissions of 1866 and 1880 proposed the separate department of agriculture in the Government of India as well as provincial levels. It was Lord Curzon who was Viceroy of India during 1901-05 whose efforts succeeded in establishing departments of agriculture at provincial levels and the North-West province UP was the first province to set up an Agriculture Department and a Central Agricultural Research constitute of PUSA] Bihar. The Indian Agriculture Service was constituted in 1906 due to the continous efforts of Lord Curzon. Why Curzon was interested in agricultural development is quite a different story. He was under the pressure of the British cotton trades who were pressurising the government for cotton cultivation in India. Thus, colonial interests were by primary objectives and rural development was secondary and implied outcome of colonial economic interest.

Second Period : (1920) to (1950) - Rural Development during National Movement

In the pre-independence era, a number of rural reconstruction steps were taken by the nationalists and social reformers during the period of the national movement. The most well-known among these attempts were the Sriniketan experiment of Tagore 1920 , Brayne's experiment 1920, Martamdum Project of Spencer Hatch 1921 , the Rural Reconstruction Project in Baroda 1932 , Firks Development Scheme in Madras 1946 , Nilokheri Experiment 1947 and the Etawah Pilot Project in U.P. 1948 .

Sriniketan Experiment

In 1920 Rabindranath Tagore laid the foundation of the Sriniketan, Institute of Rural Reconstruction and formulated a programme for the all-round improvement in the village of his zamindari with the objective of studying rural problems and of helping the villagers to develop agriculture, improving the livestock, formation of cooperatives and improving villages sanitation etc. There was marked improvement in the villages surrounding Sriniketan, but such examples were not multiplied, due to the lack of professional research support. The government was not prepared to start a full-fledged programme of rural-construction and an institution like Sriniketan had its limitations about meeting the needs of the rural communities. Tagore himself realised that the nation building work could not make further advance because it needed the combined efforts of all, a single individual or a group affort alone could not transorm the rural society. This experiment though not successful attained certain physical and notable results. The "economic returns were such that rising standard of living in the area was very noticeable. People were confident and felt able to take new things and achieve result together.

Martandam Experiment

Dr. Spencer Hatch of the Y.M.C.A. set up a project in 1921 at Martandam, 25 miles south of Trivandrum. The purpose of the experiment was to bring about a complete upward development towards a more complete and meaningful life for rural people, spiritually, mentally, physically, socially and economically. The objective of the programme was that the organisation of the Y.M.C.A. should work in the villages to eliminate poverty. The rural demonstration centre at Martandam was to guide the Young Christian Missionaries numbering 100 in Martandam and 40 in the surrounding villages. This demonstration centre had a demonstration farm, prized animals, equipment for the honey industry and other cottage vocations in its campus. Some industries like weaving, poultry, bee-keeping were started on a cooperative basis. Dr. Hatch could demonstrate the lucrative possibilities of these enterprises.

The Martandam experiment's main gains were the changes in the attitude of the rural people, inculacting in them a desire to improve, spirit of cooperation and self-respect. This experiment was followed by other states and its workers helped the State of Baroda, Mysore, Cochin and Hyderabad to set up centres for training for rural reconstruction.

Gurgaon Experiment

Mr. M.L. Brayne, who was the collector of Gurgaon District, conceived the rural development scheme in 1927. The shceme was divided into many parts such as a) institutional work; b) rural sanction; c) agricultural development; d) education; e) cooperation; f) social reforms; g) coordination and publicity. The contents of this programme were as comprehensive as those of the Integrated Rural Development Programmes. He preached the dignity of labour and self-help. Propaganda was conducted through films, songs, skits and plays with a view to increasing farm yields, and improving health standards. His great contribution was the creation of the "Villages Guide", a multipurpose worker representing the various departments of the government at the village level. Brayne's work, however, could not spread beyond Punjab and disappeared soon after he was withdrawn.

Baroda Experiment

The Maharaja of Baroda was a progressive and enlightened man and he started a rural reconstruction scheme in 1932. Mr. V.T. Krishnamachari was the Diwan of the State. He prepared and implemented a comprehensive programme of rural reconstruction, covering the various aspects of rural life to self-help and self-reliance. The programme included the items:

- Improvement of communication
- Digging of drinking water wells
- Anti-malarial measures,
- Pasture development,
- Distribution of improved seeds,
- Training in cottage crafts,
- Establishment of panchayats and cooperatives covering very village; and
- Development of village schools as centres for teaching agriculture and imparting the "will to live better".

To meet the cost of the programme a trust of Rs. 1 crore was created and in every district, intensive zones consisting of 20 to 25 villages were carved out. By 1942-43 there were 24 such centres covering 487 villages and in charge of these centres were graduate assistants to spread this message.

The Gandhian Movement

Gandhiji started his rural reconstruction activities in Sevagram near Wardha to implement his ideals of constructive programmes which included items such as the use of khadi, rural sanitation, uplift of the backwards classes, the welfare of women, education in public health and hygiene, prohibition and propagation of the mother tongue, etc. Gandhi emphasised self-sufficiency in food and cloth. He was in favour of Gram Swaraj'. His views about khadi, basic education, village self-sufficiency and the like have been actively debated by his followers and his critics. But there was agreement about his core ideas of moral values, such as truth, non-violence, non-profession, dignity of work, self-restraint, fearlessness and his insistence on the purity of the means whereby these moral ends are achieved.

The Gandhian philosophy was accepted officially and the adoption of the Khadi and Village Industry Programme, the notion of village self-sufficiency and the faith in the "Panchayati Raj" and "Sahakari Samaj" movement by the government were some of the examples of the impact of the Gandhian tradition. The Bhoodan' and Gramdan' movements led by Acharya Vinoba Bhave and Jaiprakash Naraian were the contributions as well as offshoots of Gandhian philosophy.

These early attempts were not successful because they were isolated ones except in Baroda which took the benefit of the Martandam and Sriniketan experiments; each one worked independently without taking the advantages of knowledge and experience of others. The coverage was limited due to lack of adequate resources, and because the attitude of the government of the day was unsympathetic and due to lack of technical support. However, they started the process of rural reconstruction. Tagore's approach was holistic and cultural, while Martandam's extension services were a complete inventory of project extension principles. Brayne's Scheme was comprehensive in contents. The ruling Congress party drew inspiration from Gandhiji's ideology of rural reconstruction after independence and was and still is deeply committed to rural development.

Firka Development Scheme

The Firka Development of Madras was a government launched scheme. The scheme was launched in 1946 in 34 Firkas and later it was extended to fifty Firkas[29] in (1950). The short term objectives of the scheme were to develop basic amenities and an institutional framework for carrying out communication, water supply, sanitation projects and formation of panchayats and cooperatives. The long terms obejctives were to attain self-sufficiency in food, clothing, shelter, development of agriculture, animal husbandry, khadi and cottage industries. To achieve these goals a special fund of Rs. 4 crores was creatd and at the State level, the Director of Rural Welfare was put in charge of the scheme. The Collector was made responsible for implementing the scheme at the district level. He was assisted by a Rural Welfare

Officer who was in charge of two or three Firkas and there were five to ten "Gram Sevaks' under him. Staff for agriculture and public works was also provided for every Firka.

As soon as the programme was started and implemented, the budget was diluted, originally each Firka was allotted Rs. 1,00,000 but later on it was reduced of Rs. 50,000. When the Community Development and National Extension Services Programme was adopted by Madras State in 1953-54, the Firka development shceme was merged with it.

The Nilokheri Experiment

After independence near about 7,000 displaced persons were rehabilitated in Nilokheri town; S.K. Dey, former Minister for Community Development and Cooperation was the moving spirit behind this project. The scheme was called "Mazdoor Manzil" because of its principle of "He who will not work, neither shall eat". Rights for education and medical care for the sick were guranteed. Its main point was a vocational training centre run on cooperative lines; the colony has its own dairy, poultry, piggery, printing press, engineering workshops, tannery and bone-meal factory. People were given vocational tranining of their choice to run these cooperative enterprises.

The Nilokheri project, in spite of its limitations, left an imprint on the Community Development Programme in the country. It gave the idea of agro-industrial townshp as the nerve-centre of rural development and it finds place in the First Five Year Plan and in the layout of the financial plan of the first 52 community projects.

The Etawah Pilot Project

The Etawah Pilot Project was conceived by Mr. Albert Mayer in 1948 for the development of the rural areas of Etawah district in U.P. The main objective of the project was: "To see what degree of productive and social improvement as well as initiative, self-confidence and cooperation can be developed. The problem was to ascertain how quickly these results may be attained and remain permanently a part of the people's mental, spiritual, technical equipment and outlook after the special pressure is lifted".

The other main points were production intensity, people's cooperation, development of appropriate attitudes, careful selection of personne self-reliance, local sources, and supply and development of village leadership.

In Etawah District, Maheva Block with 97 villages was selected for intensive operations. To carry out various activities like agricultural demonstrations, soil conservation, improvement in animal husbandary and village sanitation, the cooperation of various departments and non-official agencies was secured.

A programme of social education was started to secure people's participation. The main features of its organisational net were: a multipurpose worker at village level, coordination, team approach, a panel programme and regular follow-ups. The project was successful in achieving its aim and in three years it was extended to 300 villages of the Etawah district in U.P. An important contribution of Mr. Mayer in adminstrative re organisation was the practice of "Inner Demonstration", by this means warmth in interpersonal relationship, restraining arbitrary determinations and wide consultations through systematically conducted staff meetings and other adminstrative methods to bring about coordination between the technical and adminstrative personnel. In conclusion it can be said that the proejct had not only paid of the investment several times in terms of physical benefits but had also brought about non-tangible improvements of real value.

3. Third period : 1950 to 1976 (Since Independence)

After independence the framers of our Constitution tried to incorporate the main ideas thinking of the freedom movement. The dominant philosophy of our Constitution is justice in the social, economic and political walks of life. Gandhiji was emphatic about rural India and because of his inervention at the last stage, local self-government had found place in the Directive Principles of State Policy' and rural reconstruction had drawn the attention of central government and acquired a high level of priority in the Five Year Plans. The various programmes of rural development adopted by the government under our various Five Year Plans are presented in Table 1.

Table 1 - Rural Development Programmes

Five Year Plan	Programme	Year of Introduction
I	Community Development Programme	(1952)
	National Extension Service	1953
II	Khadi & Village Industries Programme	1957
	Village Housing Project Scheme	1957
	Multipurpose Tribal Development Blocks Programme	1959
	Package Programme	(1960)
	Intensive Agricultural District Programme	(1960)
III	Applied Nutrition Programme	1962
	Rural Industries Projects	1962
	Intensive Agriculture Area Programme	1964
	High Yielding Variety Programme	1966
Annual	Farmer's Training and Education Programme	1966
Plan, 1967	Well Construction Programme	1966
Annual	Rural Work Programme (RWD)	1967
Plan, 1968	Tribal Development Block	1968
Annual	Rural Manpower Programme	1969
Plan, 1968	Composite Programme for Women and Pre-School Children	1969
IV	Drought Prone Area Programme	1970

Five Year Plan	Programme	Year of Introduction
	Crash Scheme for Rural Employment	1971
	Small Farmer Development Agency (SFDA)	1971
	Tribal Area Development Programme	1972
	Pilot Projects for Tribal Development	1972
	Pilot Intensive Rural Employment Programme	1972
	Minimum Needs Programme	1972
	Command Area Development Programme	1974
V	Hill Area Development Programme	1975
	Special Livestock Production Programme	1975
	Food for Work Programme	1977
	Desert Development Programme	1977
	Whole Village Development Programme	1979
	Training Rural Youth for Self-Employment	1979
	Integrated Rural Development Programme	1979
VI	National Rural Employment Programme	1980
	Prime Minister's New 20-Point Programme	1980
	Rural Landless Employment Guarantee Programme (RLEGP)	1983
	Development of Women and Children in Rural Areas	1983
VII	Integrated Rural Energy Planning Programme	1985
	Special Livestock Breeding Programme (SLBP)	1986
VIII	Jawahar Rozgar Yojana	1989
	Prime Minister's Rozgar Yojana (PMRY)	1993
	Employment Assurance Scheme (EAS)	1993
IX	Prime Minister Gram Sadak Yojana	1997
	Rural Housing Scheme	1997
	Housing Strategies for others	1997
	Jawahar Gram Samrudhi Yojana (JGSY)	1998
	National Social Assistance Programme	1999
	Annapurna	2000
	Employment Assurance Scheme	2001
	Integrated Wasteland Development Programme	2001

Source : Prepared from various Five Year Plan documents. Planning Commission, Government of India, New Delhi.

Rural Development in Five Years Plan (First Plan to Ninth Plan)

Alleviation of rural poverty has been one of the primary objectives of planned development in India and hence rural development has been accorded a high priority in successive Five Year Plans. It is evident from the high proportion of outlay allotted to agriculture, allied activities, and community/rural development in each plan[33].

In the First Plan 1951-56, a significant programme called Community Development Programme and the national Extension Services was started to enable an integrated approach to all aspects of rural development. Emphasis was also laid on a willing and constructive participation of the people and establishment of Panchayats as civic and developmental bodies to ensure the success of the plan. The outlay in this plan for these activities was announced for 46.6 per cent of the total outlay in the public sector. This plan was not considered as development plan. It lacked comprehensiveness, certainty and coordination in its programmes and outlays.

In the Second Plan 1956-62 , the share of rural development declined to about 25 per cent which was mainly due to exclusion of outlay on power for rural development. In the first plan the outlay on power was included in the major and minor irrigation projects which formed part of rural development. Another reason for the decline was relatively high priority accorded to heavy industries. The second plan also recognised the public cooperation and participation in the scheme of democratic planning and reviewed the administration of district programmes and made recommendations. It was suggested that the general direction of policy should be to encourage local self-government and to assist them in assuming responsibility for as large a portion of administrative and social services.

In the Third Plan 1961-66 , the outlay for rural development increased narrowly. To absorb the labour force in gainful employment, the Rural Works programme was added to the general strategy of development. With its stress on self-sufficiency in food and strengthening of the infrastructure, these two segments of the economy were allocated respectively 20 and 25 per cent of the total plan outlay. It was recognised that, through democracy and widespread public participation, development along socialist lines will secure rapid economic growth and social justice. The establishment of democratic institutions at the district and block levels and the role assigned to the Gram Sabha and the Village Panchayat constitute fundamental and far-reaching changes in the structure of district administration and in the pattern of rural development. This plan, however, was a failure on a wide front. Price level rose by 36 per cent over the plan period and there was virtual stagnation in the absorption of net addition to labour force. Deficit financing was incurred on a unprecedented scale.

The Fourth Plan was delayed and the inter-regime of three years was covered by Annual Plans. During this period of break in planning, the new agricultural strategy was adopted which was marked by the beginning of Green Revolution. The Fourth Plan 1969-74 was the watershed in the development efforts. It initiated the strategy of a direct attack on poverty and unemployment. The failure of the growth oriented strategies of the sixties provided a necessary change in strategy. As a consequence, special programmes like Small Farmers, Marginal Farmers, Agricultural Labourers,

Development Scheme, Drough Prone Area Programme and Tribal Area Development Programmes for the weaker sections and economically depressed areas were introduced. Local level integrated planning was also emphasised. Panchayat Raj Institution have been established as a pattern for local development administration. In the first two years of the plan period, it was a smooth sailing due to increase in agriculture production but subsequently things went away.

The Fifth Year Plan 1974-79 recognised employment will be the most important challenge to development. Poverty was also identified as one of the basic objectives of the plan. The strategy for elimination of poverty thus rested on two major factors, a rising rate of growth of domestic product combined with declining rate of growth of population. For employment generation, the area development approach' was emphasised. To provide civic amenities and community facilities, Minimum Need Programme (MNP) was initiated. During this period, an attempt was made with respect to enactment and implementation of Panchayati Raj Act in different states. Despite all efforts, the plan could not achieve its objective and it was wounded off a year earlier.

The Sixth Plan (1980-85) has allocated substantial outlays in the public sector for plan programmes intended to alleviate poverty in the rural areas. These include programmes such as the Integrated Rural Development Programme; National Rural Employment Programme; MNP; and special programmes for Tribal areas. IRDP was launched in 2300 selected C.D. blocks and subsequently extended to all the blocks. Besides these programmes, the new 20 point programme with the objective to remove poverty was taken up. During this period, it was proposed to strengthen the process of democratic decentralisation irrespective of the existing structure pattern. Although the stipulated growth rate of 5.2 per cent per annum was achieved, there was disappointment in many front.

In the Seventh Plan (1985-90), it was proposed to achieve near full employment' and virtual elimination of poverty and illiteracy. It was aimed at to stabilise the rate of growth of the economy by taking advantage of technology to improve productivity and efficiency. The programme for rural development continued with higher priority. A need for decentralised planning at the district level and below was re-emphasised. To revitalise the Panchayats, a constitution amendment bill was brought in parliament in 1991, and the states were called upon to activate Panchayati Raj Institutions. The decline in the percentage of people below the poverty line was achieved during this period.

The central thrust of the Eight Plan (1990-95) is employment. The plan seeks to give operational content to this commitment to guarantee the right of work to every citizen through the medium of development programmes. The focus of approach under Eight Plan is : (i) Clear prioritisation of sector/projects for investment, (ii) Making available resources for these priority sectors and ensuring their effective utilisation, (iii) Creation of a social security net through employment generation, improved health care and provision of extensive education facilities and (iv) Creation of appropriate organisation and delivery system to ensure that the benefits of investment in the social sectors reach the intended beneficiaries.

In the ninth five year plan (1997-2002) includes; priority to agriculture and rural development, accelerated growth rate of economy with stable prices; ensuring food and national security for all in particular vulnerable sections of society; providing basic needs like safe drinking water, primary health care facilities, primary education and shelter; ensuring environmental sustainability of the development process through social mobilization and participation of people at all levels; empowerment of women and socially disadvantaged groups such as schedule castes, schedule tribes and other backward classes and minorities as agents of socio-economic change and development; promoting and developing people's participatory institutions like PRI's, co-operatives and self-help groups; and strengthening efforts to build self-reliance.

As briefly discussed above, it is evident that ever since the inception of planning, the policies and the programmes have been designed and redesigned with the aim to alleviate rural poverty based on increasing the productive employment opportunities in the process of growth itself. Needless to mention that the economy in this last fifty five years has progressed remarkably well but there has been disappointments in many areas. Increased participation of planning, better enforcement of land reforms and greater access to credit and inputs go a long way in providing the rural people with better prospects for economic development.

In post-independent era, government have implemented a large number of rural development programmes for the uplift of rural masses. These are as follows :

Bhoodan Movement

In (1951), Acharya Vinoba Bhave launched the Bhoodan Yojna Movement, which was a moral appeal to the land-owning classes to donate a part of their holdings for the landless workers of the society. In practice, people sometimes donated lands whose ownership was disputed or sometimes wastelands.

Community Development Programme

In (1952)), the multi-purpose Community Development Programme (CDP) was launched to bring about rural development covering all aspects, i.e., agriculture, rural industries, transport and communications, health, education, social welfare of women and children. The central idea behind this programme was to develop self-reliancte and self-help both in the individual and the community to achieve the goals. According to the First Five Year Plan : "Community Development is the method and rural extension is the agency through which the Five Year Plan seeks to initiate a process of transformation of the social and economic life of the villages."

This programme was initially started by the Government of India by entering into an aid agreement with the Ford Foundation to start 52 Community Projects in areas endowed with irrigation facilities or assured rainfall. In the Third Five-Year Plan, it was gradually extended to cover the entire country. This programme has helped in setting up a network of basic extension and development services in the rural areas and introduced administrative machinery, which formed the basis for planning and implementing rural development programmes. CDP did not yield the expected results because there was no institutional mechanism for the local people to participate in the development work.

This programme also could not bring about a significant increase in agricultural production either because wide range of activities were to be covered within limited resources and lack of co-ordination at the district, block and village levels. As a result, the country faced acute food shortage, which pressurised the government to import huge quantity of food-grains and to increase the food-grains production as well.

Intensive Agricultural District Programme

Intensive Agricultural District Programme (IADP) was launched in (1960)-61) in 15 selected districts] one district in each state . The idea was given by the Ford Foundation in (1959) in its recommendations on the Food Crisis' report. The Government of India approved and included IADP in the Third Five Year Plan, which stated: "In pursuance of this proposal, IADP has been taken up, to begin with, in one district in each State. The programme is intended to contribute both to rapid increase in agricultural production in selected areas and to suggest new innovations and combinations of practices which may be of value in other areas".

This programme was not established as a separate programme as it used the community development organisation and operated through the Community Development Blocks. It suffered from various limitations and severe droughts in 1965-66 and 1966-67. But its concept of package of improved practices evolved for individual crops, consisting of the use of improved seeds, proper soil and water management, fertiliser and pesticides etc., was found useful and practicable. Therefore, the government decided to expand it to other districts through Intensive Agricultural Areas Programme.

Intensive Agricultural Areas Programme

Intensive Agricultural Areas Programme (IAAP) , launched in 1964-65, was a modification of IADP. It was however, slightly different from the IADP as it emphasized intensive agricultural development of the area as a whole instead of districts on the basis of potential of a single crop. In the country, it was for the first time that a sizeable proportion of cropped area was brought under intensive production but without increase in resources. The Committee of Experts described the programme as a "path finder" and "pace setter" for the whole agricultural development programme, consisting of three-tier system in which they envisaged IADP as the first tier, the IAAP as the middle tier and Normal Development Programme as the lowest tier. This paved the way for launching the High-Yielding Variety Programme.

High Yielding Variety Programme

In 1965, the adoption of high-yielding variety of seeds backed by fertilizers, pesticides, adequate water supply, agricultural machinery and other equipment marked a turning point in the modernization of traditional agricultural system. HYVP became the instrument of agricultural growth. The draft outline of the Fourth Five Year Plan, stated: "It is necessary to make far greater use of modern methods of production and to bridge the gap between demand and production by the application of the latest advance in the science of agriculture".

This program suffered because of drought conditions in many parts of the country. The major limitations in the program were its bypassing such regions, which were not endowed with, assured irrigation facilities or sufficient rainfall and the rich farmers derived the benefits of the new technology.

Small Farmer's Development Agency and Marginal Farmers and Agricultural Labourers Projects

The "Garibi Hatao", Poverty Eradication Policy of Mrs. Gandhi's Government became the main thrust of the Fourth five-year Plan under which a whole lot of program for the development of weaker sections of society were launched. The plight of small and marginal farmers attracted the planners and experts after the All India Rural Credit Review Committee drew their attention to the highly unequal distribution of rural credit to big farmers and the exclusion of small and marginal farmers by the credit agencies. It recommended the formation of the Small Farmer's Development Agency, (SFDA) to help small farmers. As a result, in 1970-71, the government started 46 pilot projects for small farmers and 41 projects for marginal farmers and agricultural labourers in selected districts.

These projects did not have their own field staff. To fulfil their objectives, i.e., identification of small farmers, marginal farmers, agricultural labourers and their problems, formulation and execution of model plans to solve the different problems and to review the progress of the execution of these activities as well as the effectiveness of the efforts undertaken to benefit the target groups, they were given necessary assistance by the existing agencies and financial institutions.

However, SFDA and MFAL projects, designed to raise the income level of target groups of the agency, resulted in despair because of the lack of precise criteria, delay in the formation of schemes, faulty land records, lack of their own field staff, apathy of commercial banks to finance the schemes, and misuse of subsidy. It was then realized that the problems of these weaker sections cannot be solved by economic measures alone. Thus, for the overall development of the concerned target groups, these special programmes were merged with Integrated Rural Development Programme.

Drought Prone Area Programme

Indian agriculture mostly depends on the monsoon. Inadequate rains result in drought. As a consequence, scarcities of food and drinking water badly affect the lives of rural poor. To restore an adequate ecological balance, the Drought Prone Area Programme, (DPAP) was launched in 1970.

This programme is primarily based on the preparation of district development plans for optimum utilization of land, water, livestock and human resources. Agricultural labourers and small farmers require regular employment to earn their livelihood. However, DPAP seeks to stabilize the income of the poorest segments of rural society.

Desert Development Programme

In 1977-78, the Desert Development Programme (DDP) was introduced in the arid desert areas of Rajasthan, Gujarat, Haryana, Himachal Pradesh and Jammu and kashmir. The main thrust of the effort is directed towards checking desertification through activities which restore the ecological balance, providing facility for soil and water conservation and stabilizing sand dunes. The efforts have been made to achieve higher income and employment opportunities for the local residents by raising the productivity and productive resources of the area. A significant success has been achieved in the scheme of afforestation, animal husbandry, drainage and rural electrification.

Command Area Development Programme

In 1978 the Command Area Development Programme (CADP) was launched for the maximum utilization of the irrigation potential. The efforts and investments made earlier created irrigation potential to some extent, but its utilization remained inadequate for want of land consolidation, construction of field channels to carry water to individual fields, drainage system and roads for carrying products to the market by the farmers.

Crash Scheme for Rural Employment

The Crash Scheme for Rural Employment (CSRE) , introduced in 1971, was meant to provide employment during the working period of ten months in a year to about 1,000 individuals in every district. It also emphasized the creation of assets of a durable nature as per local development plans. Under CSRE, three important schemes adopted included land development, minor irrigation and road construction. The operation of this programme revealed various shortcomings. The basic limitations in this programme were lack of coordination among various departments and conflict with the State plan programmes, e.g., mostly the projects were listed and identified after the sanction of funds and without any reference to the preparedness of Panchayati Raj agencies responsible for carrying out the projects. The crash programmes proved much wasteful and failed to create a stable and permanent base for employment in rural areas.

Minimum Needs Programme

Minimum Needs Programme (MNP) was started in 1972 with the main objective to raise substantially per capita consumption of the rural population living below the poverty line. Its main thrust is on providing supply of drinking water, nutrition, health facilities, elementary education, roads, houses for rural landless and electrification in the rural areas. It is basically a human resource development programme. The implementation of the programme in the States of Haryana, Punjab, Karnataka, Andhra Pradesh, Tamil Nadu, Maharashtra and Gujarat proved successful. But in Bihar, Madhya Pradesh, Orissa, West Bengal and north-eastern States, it lagged behind due to lack of co-ordination between the Centre and the States and between the district authorities and block agencies.

Food for Work Programme

The Food for Work Programme (FWP) was introduced in 1977, to generate additional employment in rural areas through projects designed to create durable community assets to strengthen the rural infrastructure. The labourers were paid in terms of foodgrains for the work done by them. This was made possible due to the availability of surplus foodgrains in the country. In the implementation of this programme, PRIs played an important role. This programme achieved a great success in the short period of the first three years as it created substantial additional standard of the rural poor by providing them foodgrains at cheap rates. However, the programme suffered from various shortcomings at the level of planning and supervision. In August 1980, the programme was renamed as the National Rural Employment Programme and became an integral part of the Sixth Five-Year Plan.

Integrated Rural Development Programme

Integrated Rural Development Programme (IRDP) was launched in 1978-79. It is a major instrument to wipe out rural poverty. Its main objective is poverty alleviation through growth and generation of employment opportunities for the poorest of the poor in rural India. The benefits were to reach to the identified target group comprising of scheduled castes, scheduled tribes, small and marginal farmers, tenants, landless labourers, share-croppers and rural artisans. This programme is implemented through District Rural Development Agencies. For the implementation of programme at the grassroot level, the block staff is responsible. IRDP, a programme of massive dimensions, having a multiplicity of critical parameters and functioning in a highly diverse environment have achieved a limited success.

Training of Rural Youth for Self-Employment

TRYSEM is a facilitating component of the Integrated Rural Development Programme, which was launched on 15th August, 1979. The objective of TRYSEM is to provide technical and managerial skills to the rural youth in the age group of 18-35 from families living below the poverty line to enable them to take up self-employment ventures in the broad fields of agriculture and allied activities, industries, services and business. Priority is given to rural youth from scheduled castes and scheduled tribes in the programme. Their minimum coverage should be 30 per cent. The coverage of women should be at least one-third of the total number of youth.The training under TRYSEM has been provided to about 6.9 lakh rural youth from families below the poverty line during the first four years of the Seventh Five Year Plan.

National Rural Employment Programme

The National Rural Employment Programme (NREP) was started in 1980. Expenditure on this programme was shared on fifty-fifty basis between the Centre and the States. NREP is the culmination of experiences in implementing employment programmes in the previous years, i.e., Crash Scheme for Rural Employment, Pilot Intensive Rural Employment Projects, Food for Work Programme, etc. NREP has been merged into Jawahar Rozgar Yojana in 1989. The basic objectives of the programme were to generate additional employment opportunities; to create durable community assets; and improvement in quality of life of rural masses[52]. The implementation of this programme has resulted in providing great relief to rural people and facilities for trade and commerce with the improvement in communication. It has ensured minimum wages.

Rural Landless Employment Guarantee Programme

RLEGP was launched on the 15th August, 1983 with the basic objective of generating gainful employment for the rural landless labourers by creating productive assets for strenthening the rural socio-economic infrastructure and improving the overall quality of life in the rural areas. Due to paucity of resources, the guarantee part of the programme could not be implemented. The priority in employment was given to landless labourers, women, scheduled castes and scheduled tribes. Under RLEGP, funds were earmarked for Social Forestry, Indira Awas Yojana, and construction of Rural Sanitary Latrines.

Development of Women and Children in Rural Areas

DWCRA programme was launched as a pilot project in 50 selected districts in all States in 1982-83. On 31st March 1988, DWCRA was being implemented in 106 districts. It was formulated as a sub-scheme of the IRDP with focus on the rural poor women to provide them with suitable avenues of income generation according to their skill and local conditions.

The rationale behind these schemes is that women's incomes have positive co-relation with the nutritional and educational status of the family and in the building up of a positive attitude towards the status of women. The financial provisions under DWCRA are meant for the groups of women only. An amount of Rs. 15,000 is given to each group of 15-20 women, which can be used for purchasing the raw materials, marketing and child-care facilities. This programme is implemented through the District Rural Development Agencies. Since its inception, more than 3,79,641 rural women below the poverty line have been assisted through this programme and about 22,682 groups of women have been formed to take up income generating activities[57].

Twenty-Point Programme

In 1975, the then Prime Minister, Indira Gandhi initiated the first 20-Point Programme to mount an attack on poverty. The changes in the socio-economic conditions, fulfillment of several targets and the new challenges made it necessary to revise this targets and the new challenges made it necessary to revise this programme and the revised 20-Point Programme was announced on 14th January, 1982. This programme played a significant role in uplifting 10 crore people above the poverty line, but 1983-84 estimates show that still 27.1 crore of people live below the poverty line. The programme launched on 1st April, 1987 along with the Annual Plan of 1987-88, has its thrust on removal of gross anomalies of the system, which could not bring all the people above the poverty line in spite of sizeable achievements in agriculture, industry and technological sectors of Indian economy.

The programme's coverage has been broadened to bring within its fold a number of major areas of social concern, such as special programme for rural labour, provision of clean drinking water, health facilities for all, two-child norm, expansion of education, equality for women, new opportunities for youth, justice to scheduled castes and scheduled tribes, housing for people, slum improvement, protection of environment, and concern for the consumer.

Accelerated Rural Water Supply Programme (ARWSP)

The aim of ARWSP is to provide adequate and safe drinking water facilities to the rural people by supplementing the efforts made by the State Government under the Minimum Needs Programme. To cover all problem villages by 1990, the National Drinking Water Mission was launched in 1986, to facilitate scientific source, find and initiate steps to improve the quality of water under specific sub-missions on guinea-worm eradication, removal of excess fluoride, iron and salinity. Satellite imageries and geophysical investigations have been helpful in locating sources in problem villages.

Jawahar Rozgar Yojana

The existing rural wage employment programmes - NREP and RLEGP got merged into a single scheme Jawahar Rozgar Yojana, announced by the then Prime Minister, Rajiv Gandhi in the Parliament on 28th April, 1989.

The scheme aims at reaching every Panchayat and seeks to provide employment to at least one person in a family living below the poverty line in the rural areas[61] for 90 to 100 days a year at a place near his residence. Nearly 440 lakh families are living below the poverty line in rural areas. In all, 30 per cent of the jobs will be reserved for women and preference will be given to scheduled castes/scheduled tribes in employment.

The resources for the scheme are allocated to the States/Union Territories on rural poverty incidence. The entitlement of funds to a Gram Panchayat depends on the per capita allocation of the district worked out on backwardness criteria given below as well as population. The bases of district allocation of funds are: (a) rural scheduled castes/scheduled tribes population as per cent of total rural population (60% weightage), (b) agricultural workers as per cent of main workers (20% weightage), and (c) inverse of agricultural productivity - per hectare value of agricultural produce (20% weightage).

Prime Ministers Gram Sadak Yojana

There are more than 2.5 lakh villages, which are not connected by roads. There fore for the first time in Indian history the scheme of building roads came to form. The objective of the scheme is to connect all villages having the population of 1000 by the year 2003 and villages having 500 population shall be connected by the year 2007 so that nearly the population of 1 lakh will be benefited by the roads. The other objective is to repair the existing rural roads (5 lakhs KM) . The total expenditure will be provided by central government.

Rural Housing Scheme

Shelter being the fundamental right of every human being, central government provides a house for everyone below poverty line. To eradicate homelessness is the objective of the current Ninth Plan. By the year 2007 all the kaccha houses will be made pucca (cement & bricks) .

Housing Strategy for Others

For the population falling in the income less than 3,200 annually can avail the loan through commercial banks, housing boards and other financial institutions upto Rs. 40,000 and Rs 10,000 will be provided as assistance from the government. These schemes will be implemented by DRDA or Zilla Parishads. For the rural population and people below poverty line also have several loan schemes of HUDCO.

Jawahar Gram Samrudhi Yojana (JGSY)

This programme was started in the year 1999. The aim of this programme is clean, independent and self-sustaining villages. It has provided employment opportunities to the rural poor. Gram Panchayat with the consent of Gram Sabha implements this programme. The financial assistance is received from central government, DRDA and Zilla Parishad. 22.5 per cent of these funds will be allotted for the schedule caste and schedule tribe.

National Social Assistance Programme

This programme was initiated for the old and destitutes. This involves the national plan for old age pension and national family welfare plan.

Annapurna

This schenme is for the people below poverty line. The government provides grains at minimum cost. The pensioners above 65 years of age below poverty line get 10 kilos of grains free of cost. The beneficiaries are selected by the Gram Sabha and allotted the cards.

Employment Assurance Scheme

This scheme is for the rural poor below poverty line during the time of famine and the period of no income or no employment. This programme has been implemented throughout the country.

Integrated Wasteland Development Programmes

To increase the production capacity of the wasteland and to raise the standard of the rural poor central government has initiated this programme. 100 per cent of assistance is provided by the central government by implementing watershed programmes.

Swarnajayanti Gram Swarozgar Yojana (SGSY)

The governments SGSY has been conceived of as a mixed bag of government credit cum subsidy along with the concept of self-help as promoted by the NGOs to meet the target of 30 percent of the people living below poverty line. With SGSY in operation, the earlier poverty alleviation and income generation programmes- IRDP, TRYSEM, DWCRA and MWS will no longer be in operation. SGSY is supposed to be a holistic programme, conceived for convergence of various schemes under one umbrella. The aim of the government scheme is to promote self-help groups of people living below poverty line, and to promote the establishment of micro-enterprises in rural areas, which ought to yield a monthly income of Rs. 2,000 per family within three years.

Problems of Rural Development

As noted earlier, India continues to face the problems of unemployment and poverty even after five decades of development efforts by the independent government. The complexities and ill-effects of the past development models call for a multi-dimensional, people-centered rural development to meet the diversity in poverty and poor and marginalised groups. The government agencies and political parties failed to bring in people-centered rural development. In this context, Webster states that:

- A government rarely does, or will fulfil more than a few of the wide range of demands that effective democratic decentralized government requires. Rarely, if ever, is there an adequate devolution of power, of responsibilities or of resources to decentralized government institutions.
- Secondly, the government is rarely willing to implement the types of structural reforms and policies that can bring about a transformation in the abilities of marginalised groups and other disadvantaged social factors to contest more successfully in key markets that determine the economic, social and political conditions.
- Thirdly, institutions of local government are rarely willing to bring about the mobilisation of disadvantaged groups in order to place demands upon the State.
- Fourthly, the electoral focus of political parties upon the institutions of government at the local and national levels tends to mitigate against taking up specific local problems or, given the patrimonial nature of local politics, problems that challenge local political elites.
- The various programmes and approaches that have been examined confirm that no single package or formula is sufficient for effective rural development. The major weakness of all these approaches are:
- The approaches mainly concern with agricultural development and focus is individual cultivator;
- Restructuring the administrative apparatus at the district and below (block level) to suit the bureaucracy;

- The approaches concern the rural poor but their implementation failed to eradicate poverty and identification of the rural poor ;
- Some of the services and programs like Community Development Program , Command Area Development and Integrated Rural Development, etc., are the creation of foreign experts and agencies. The financial support along with these programs immediately draws our attention and these schemes condition the minds of our administrators, policy-makers and academicians;
- In all the approaches funds would not be directly given to bodies at the grassroots level and before the funds are spent, they percolate down at least through various layers of the administrative machinery and because of many intermediaries there are losses of various types. The rural people have almost no control over the funds, which are supposed to be spent on them by the government, and they depend on bureaucracy for the benefits to trickle down. Thus, self-respect and confidence have been victims of this system.
- Mere extension of past concepts and practices which are away from reality would not serve the purpose and major re-thinking is required to develop the rural India and the whole handling calls for adaptation, modification and experimentation, depending upon the urgent attention to the situation in our country.
- The goal of rural development is to remove the constraining rural characteristics of society which are:
- A large percentage of the population depends on agriculture;
- There is a low purchasing capacity of farmers;
- There is strict control over rural institutions by State bureaucracy; and
- There is a lack of strong economic organisations of rural people.

Emergence of Voluntary Agencies

Due to the failure and weakness of government policies for rural development the NGOs come to act as an intermediary development agencies between the State and the target groups by acting as their guide, philosopher and friend. These agencies could play this type of role because they have distinct qualities like dedication, innovative approach, local area's knowledge, rapport with the people at the grassroots, middle and higher levels, non-rigid approach, effective motivation and almost no procedural implications, which are rarely found among the governmental functionaries.

The NGOs became prominent after independence, especially after 1970s. This was partly because of the limited success of past development policies pursued by the government. Even after half-a-century of development efforts initiated by the State, the problems of largest concentration of the poor, hunger, malnutrition, unemployment, gender inequality, illiteracy, etc., continue to plague Indian Society. One of the important contributing factors for the limited success of rural development programme was the absence of involvement of the people for whom the programmes were meant. The need for micro-level institutions to involve the people in formulation, implementation and monitoring of the programmes is, therefore, stressed in several quarters. Development practitioners, government officials and foreign donors consider that Non-governmental Organisations (NGOs), by virtue of being small scale, flexible, innovative and participatory, are more successful in reaching the poor and in poverty alleviation.

This consideration has resulted in the rapid growth of NGOs involved in initiating and implementing rural development programs. Scholars and activists working in the NGO sector have arrived at an estimate of about 30,000 NGOs in India. A rapid growth of NGOs took place in the 1980s and the early 1990s.

Today there are about 31 lack registered NGOs in India as per the study by CBI. (ref. Indian Express 1 Aug, 2015).

The term NGOs' is used to specify those organisations, which undertake voluntary action, social action and social movements. The term is negative in the sense it seeks to give a meaning that NGOs possess the features not possessed by the government, and undertake activities otherwise normally not undertaken by the government. The following four characteristics make the NGOs as distinct organisations; voluntary formation, working towards development and amelioration of suffering, working with non-self serving aims and relative independence. NGOs are voluntarily formed in the sense that there is no compulsion from government or others, which leads to their formation. There is also an element of sacrifice in the fact that the staff, especially at leadership levels, works at salaries below what they can draw in the government or the private sector. This, however, may not apply to funding agencies or to professional NGO where salaries paid to the staff members may be higher than those paid in the corporate sector. They are development-oriented in the sense that they are concerned with improving the condition and position of oppressed sections of society, as opposed to other goals like entertainment, promotion of religion, etc.

Finally, they are relatively independent from the government in the sense that their Board of Directors or Trustees determines their policies. However, the NGOs have to work within the parameters of government legislation and policies formulated for NGOs.

Types of voluntary agencies and their functions: NGOs can be classified under four broad categories:

- Operational or grassroots NGOs
- Support NGOs
- Network NGOs
- Funding NGOs

Operational or Grassroots NGOs

Grassroots NGOs directly work with the oppressed sections of society. Some NGOs are big, while some are small. The grassroots NGOs could be either local based, working in a single and small project location, or be working in multiple project areas in different districts, States and regions covering a larger population.

The approach and orientation of grassroots NGOs also differ. Based on this, the following distinction can be made among grassroots NGOs: (i) charity and welfare NGOs' focus on proving charity and welfare to the poor, (ii) development NGOs' focus on implementation of concrete development activities, (iii) social action groups' focus on mobilizing marginalised sections around specific issues which challenge the distribution of power and resources in society and (iv) empowerment NGOs combine development activities with issue based struggles.

Charity and welfare NGOs are involved in charity giving food, clothing, medicine, alms in cash and kind, etc., welfare] providing facilities for education, health, drinking water, etc., relief responding to natural calamities like floods, drought, earthquakes and man-made calamities like refugee influx, ravages of war, etc. and rehabilitation undertaking the work in areas struck by calamities and starting activities durable in nature. A large number of church based NGOs operating in south and northeast India still have charity and welfare component in their program.

Development NGOs may be involved in facilitating the provision of development services such as credit, seeds, fertilizers, technical know-how, etc. Such NGOs concentrate on development of socio-economic environment of human beings.

Social action groups focus on mobilising marginalised sections around specific issues, which challenge the distribution of power and resources in a society. These NGOs are involved in raising of consciousness of the people, awakening, organising, recording of priorities to suit social justice, redeeming the past and opening doors of opportunities to the oppressed and the exploited. Young India Project (YIP) , having office at Penugonda, A.P., has been involved in mobilisation has mobilized the people for an effective implementation of land reforms and for the introduction of Employment Guarantee Act, They also enable the poor to stage dharnas, protests to obtain government programs, etc.

Empowerment NGOs combine development activities with issue based struggles. They may be involved in the provision of services such as savings and credit; but they utilise such activities for social, economic, political and cultural empowerment of the poor. MYRADA utilizes credit management groups, and watershed programs for not only to bring development among the oppressed communities but also for social and political empowerment.

The main difference between social action groups and empowerment NGOs is that the former do not normally undertake development activities as they believe in addressing the root causes of poverty. On the other hand, the empowerment NGOs undertakes development activities because the people cannot undertake the struggles with empty stomachs'. They believe that empowerment of the people is an essential pre-requisite for development. Hence, they strive hard to enable the people to become free from all the exploitative structures.

The approaches followed by the charity, welfare and development NGOs are related to delivery system. As the activities undertaken by first two types of NGOs are non-controversial and do not lead to clash of interests in the countryside, the government too extends full support to NGOs working in these areas. The support from donor agencies to NGOs involved only in charity and welfare is on the decline as it is felt that this strategy is based on giver and receiver' relationship and not that of building the capacity of the people. It is also believed that it is paternalistic in nature and cause human degradation.

With the increase government funding for anti-poverty programs through NGOs and the growing legitimacy for NGOs, many government officials and political leaders also joined the fray often by floating their own NGOs. Grassroots NGOs now undertake a host of activities including environmental projects, dryland development, savings and credit programs, schemes for income generation, health and education projects, the formation of agricultural labour unions, etc.

Support NGOs

Support NGOs provide services that would strengthen the capacities of grassroots NGOs, Panchayati Raj Institutions, cooperatives and others to function more effectively. Examples of this type of NGOs are SOSVA, SEARCH, etc. Some do not engage in grassroots action while others do have field projects, but grassroots action is not their primary task. Some of the support NGOs render support in specific thematic areas such as health, education and environment, while others provide support in generic issues such as perspectives, leadership, human resources development, management, etc. The support NGOs bring out periodicals and run training programmes for the activities involved in NGOs, Panchayati Raj Institutions, etc. For this, they have set up training institutes with facilities to train a number of participants.

Umbrella or Network NGOs

Network NGOs are formal associations or informal groups of grassroots and/ or support NGOs, which meet periodically on particular concerns: An example to this is FEVORD-K (Federation of Voluntary Organisations in Karnataka). They act as a forum to share experiences, carry out joint development endeavour as well as engage in lobbying and advocacy. The participation of network NGOs in lobbying and advocacy is, however, a recent phenomenon.

Funding NGOs

The primary activity of these NGOs is funding grassroots NGOs, support NGOs or people's organisations. Most funding NGOs in India generate a major part of their resources from foreign sources, though there is an effort by some to raise funds from within India. The organisations such as CRY, Dorabji Tata Trust, Aga Khan Foundation, in India provides funds to NGOs. Foreign NGOs like NOVIB (Netherlands International Development Cooperation), Action Aid, Oxfam, etc., with headquarters in the developed western countries, mobilize resources from both the public and governments in their respective countries to help grassroots NGOs in their efforts to initiate and implement pro-poor rural development activities. Foreign NGOs do have field offices here. Some bilateral agencies like German Development Corporation (GTZ), DANIDA, Canadian International Development

Agency (CIDA), etc., do provide funding support to NGOs, but, these cannot be called as NGOs.

The voluntary action in different parts of India was ruled in a specific socio-political context and was inspired by the emergence and continuity of social reforms, social change and political movements in different parts of the Country. All along this period voluntary action had also matured and began to emerge in extension and conscientization and organization of people. What started as social work with a focus on charity has veered towards developmental work and community mobilization. There has been a proliferation of voluntary organisations across the board, each with its specific perspectives, priorities and strategies.

Role of Voluntary Organisations in Rural Development

The voluntary sector plays an important role in development particularly with reference to the programmes aimed at the rural poor and for improving the quality of life in rural areas. The Ministry of Rural Areas and Employment has been earmarking funds for the voluntary sector not only in the nationally planned programmes but also for mobilizing and motivating people's participation. With a view to encouraging, promoting and assisting voluntary action in rural development and with focus on injecting new technological inputs for the enhancement of rural prosperity, the Government of India, in September 1986, set up the Council for Advancement of People's Action and Rural Technology (CAPART), a registered society under the aegis of the Ministry of Rural Areas and Employment by merging two autonomous bodies, namely People's Action for Development India (PADI) and Council for Advancement of Rural Technology (CART). The Cabinet Minister /Minister of State incharge of the Ministry of Rural Areas and Employment is the President of the General Body of CAPART. The General Body comprises officials of the Ministry of Rural Areas and Employment, eminent social activists, experts in different fields in rural development and representatives of voluntary agencies. The General Body lays down the broad policy framework within which projects under various schemes are sanctioned.

Capart's Assistance

While Capart seeks to associate the voluntary sector in several schemes that are part of planned development and are being implemented on national basis, it also supports several innovative projects keeping in view the needs of specific areas. The main thrust of these programmes is on employment generation, houses for shelterless, conservation of water, enhancement of income and development of community assets. While Capart receives most of its funds from the Ministry of Rural Areas and Employment, it also channelises some funds of VOs from funding agencies like DANDIA, SDC, etc. Efforts are also being made to mobilize international donations both from multilateral agencies and other international donors. A beginning has been made in this direction.

Capart provides assistance to VOs under different programmes like Indira Awas Yojana, Watershed Management, Social Forestry, Link Road in hilly areas, Promotion of Voluntary Action in Rural Development, Development of Woman and Children in Rural Areas, Organisation of Beneficiaries, Central Rural Sanitation Programme, Drinking water under the Technology Mission on Accelerated Rural Water Supply Programme and Advancement of Rural Technology. Since 1995, Capart is also providing assistance for talking up capacity building and rehabilitation programme for persons with disability. 3% of the allocations made under different programmes of Capart are being set aside to fund this programme. It has also been decided to give preference in the coverage to disabled beneficiaries under the programmes.

Promotional Role

Capart has a significant role in promoting a variety of activities of transfer of technology, people's participation, development of marketing for products of rural enterprises, generating awareness among rural masses for bringing about attitudinal changes towards persons with disability to contribute in the field of rural development and promotion of other developmental activities and delivery systems in non-governmental sector.

Examples of such activities are as follows :

Housing Development

Housing programme encourages not only economic activities but also generates employment opportunities and creates a solid base for healthy and hygienic living. Capart has been providing funds to VOs for construction of houses with infrastructure such as roads, soakpits, other common facilities and individual sanitation units to the targeted population consisting of SCs/STs/freed bonded labourers and other socially and economically weaker sections of the society through Indira Awas Yojana.

Watershed Management

In a major attempt to tackle the ecological crisis and to conserve water to bring about all round development with the Government announcement of Integrated Watershed Conservation and Development Programme, own or due to efforts of a Voluntary Organisation for a cause which is sufficiently just or serious and to sustain their campaign/struggle for their betterment of economic status or social power.

The role of NGOs in the rural development is highlighted in the following.

Developing Group Entrepreneurship among IRDP Beneficiaries

IRDP beneficiaries are those have no accesses to market and financial and other infrastructural facilities. They live below the poverty line. These beneficiaries should be organised into cooperation with the support of NGOs at the village level so that they can market their products at reasonable prices. The IRDP beneficiaries are more in traditional activities like dairy farming, business shops, bullock carts and repairing activities. These beneficiaries should be organized into co- operatives with the support of NGOs at the village level so that they could secure reasonable prices for their products. The NGOs can help at the grass root level to motivate beneficiaries to change their occupations from traditional activities to Industry Service Business (ISB) Sector.

Organisation of Rural Poor

The rural poor may be organised in the following manner.

Cooperatives of agricultural labour : If some land is given to them, they may manage it properly and make it a viable unit of operation.

Cooperatives of artisans : Lack of facilities for procuring raw material and marketing of finished goods and the stumbling blocks for the upliftment of these artisan cooperatives are set up to improve their standard of life.

Cooperatives of Marginal Farmers : The marginal farmers hold land, which is not economically viable, or they don't have sufficient means of cultivation. By organising them on cooperative basis, they shall get the economics of scale, which have been so far enjoyed by big farmers.

Cooperatives of Dairy farmers : The individual dairy farmers may be organised on cooperative basis so that they can get reasonable prices of their products.

The organization of rural people can only be practical proposition provided there is the involvement of the NGOs particularly at the initiation stage.

Implementation of Jawahar Rozgar Yojna (JRY)

JRY is a program of the people, by the people for the people. The gram panchayats are responsible for planning and execution of this scheme. This is the largest program. The primary objective is to greater gainful employment for the unemployed, underemployed men and women in rural areas. Secondly to create sustained employment by strengthening the rural people. The PRI's are not resourceful to plan and execute this yojana independently. The assistance of NGOs is required for desired results. The involvement of NGOs is required at the time of selection and execution of projects. Thus they can play a vital role in the proper implementation of the JRY by creating the awareness among the rural people

Interaction among Volunteers, Technicians and Scientists for Rural Development

The interaction among volunteers, technicians and scientists is very essential to utilize local human and natural resources in a optional way for the development of local areas. In this regard the reference of AMUL Cooperative of Dist. Khera in Gujarat is a pertinent example. A volunteer Shri T.S.Patil leed this cooperative and technocrat Dr.Kurien provided technical know-show and now the Amul cooperative dairy has spread throughout Gujarat. The NGOs should promote these types of innovative projects at village, block and district levels.

Creating Community Groups

Bal dal, Yuvak dal, Kisan dal, Mazdoor dal, and Vidyarti dal may be created at the village level for doing some constructive work on behalf of the village community. These dals require professional guidance to carry out various development activities. The NGOs can help these institutions by way of providing technical/professional help in carrying out the projects. It can identify dedicated youths, kisans, mazdoors who have leadership qualities, this should be inducted in the respective dals.

Promoting the Non-conventional Sources of Energy and Conservation of energy

There is a need of an appropriate database for the demand and supply of the energy in rural sector. There should be an effective planning system to use the available source of energy and management of existing energy sources. A very large potential exists in biogas, wind, biomass and solar source of energy, which are currently being poorly tapped. The NGOs should come forward and encourage the people to generate more non-conventional source of energy and use the same effectively in the production function.

Integration of Schemes

To alleviate poverty and provide minimum needs to rural population many government departments like rural development, health, education etc., are providing benefits to rural people separately without knowing the efforts being made by each other. Thus there is a need to co-ordinate various schemes of different departments at the village level by the NGOs acting as a catalyst. The NGOs can also initiate (with the help of Panchayats) to establish village secretariat where the personnel of various departments may come once or twice in a month to apprise themselves as well as people about the latest programs of their department.

Disaster Relief

Over the last two decades India has faced serious large-scale natural disasters like floods, droughts, cyclones and earthquakes. NGO's have done a lot work in this field ranging from short-term relief to long-term rehabilitation and development.

In India the types of roles and responses to situations of natural disasters by the NGO's can be broadly classified into the following categories :

- For large relief agencies and NGO's, the main response are to provide material relief and rescue operations during times of disaster including medical relief.
- This followed by a longer period of reconstruction activities of the physical infrastructure like roads, houses, community buildings, drinking water facilities etc., and continuation of medical aid.
- For the small and localized NGO's, the initial response in the form of rescue and material relief reconstruction of houses are also undertaken.
- Most of the larger India Agencies, which are not located in the disaster prone areas, withdrawn after the initial phases of relief and reconstruction, while only a few prolong their presence in the areas for restarting some development activities.
- Local NGO's, participate in relief and reconstruction activities during disaster.

- **Drought Relief:** According to National Commission of Agriculture (1976) drought are three types:
- Meteorological Drought
- Hydrological Drought
- Agricultural Drought.

About 33 percent of the country is severely drought prone. About 35 percent of the country's cultivable area is drought prone because of the erratic rainfall. Shortage of rainfall should not solely be considered as the only basis for drought intensities. Several other factors such as temperature wind velocity, soil texture, ecotranspiration, and stage of crop growth and antecedent rainfall all interact to produce drought situations.

NGOs work in various ways like

- Promotion of microshed around water harvesting structures. The activities of which are conducting pilot projects. Identification of rainwater harvesting with ponds, percolation tanks, check dams, including renovation of existing water. Skill training in technical aspects of micro watershed.
- Dry land agriculture and appropriate land and water use. Promotion of dry land agriculture projects. Identification and demonstration of drought resistant varieties of crops. Training is given low cost techniques on land developing, catchment treatment, soil and moisture conservation structures.
- Identification and strengthening of Local Croping Mechanism.
- To facilitate better coordination between NGO and government for drought related programmes.

Rural Women and Child Health Programmes, Family Planning Programmes

Maternal mortality is an index of the socio-economic development and standard of health care especially in the developing countries. About 600,000 women die every year from complications of pregnancy and childbirth. India alone accounts for 25 percent of such deaths. For every one death between 30 to 100 women suffer from acute maternal morbidities which are painful debilitating and often permanently disabling.

The global Safe Motherhood Initiative (SMI) was launched at an international conference in Nairobi Kenya in 1987. Making pregnancy safer is not only a health issue it is also a moral issue. It is essential that all pregnant women have access to quality obstetric care, which is effective, accessible and acceptable to them. The role of NGO's in this SM programme is :

- Applying the provision of human rights instrument - to ensure SM by fulfilling the rights of girl and women especially to participation, education, nutrition, health care, safe environment, to function for discrimination, and protection from violence and abuse.

- Encouraging the governments to make sustained social investments - higher investment in health care including pre and postnatal care, prenatal care, emergency obstetric care, skilled attendance at delivery, Family planning, reducing teenaged pregnancy, prevention and care for HIV / AIDS infected, HIV / AIDS awareness to women.
- Help to establish women friendly health services
- Helping communities to support women
- Encouraging the formation of women's groups

Involvement of NGO's and community participation in the planning design and implementation of Safe Motherhood Programmes, better transport and communication systems, training village local midwives user friendly health services and better referral systems are recognised as critical intervention.

NGO and the Indian Rural Labour Markets

Rural labour markets play very important role in rural development and there we find the example of mobilization of people for their rights. The Institutional factors, which determine the land productivity and labour productivity, are played by many variables. Conventionally beside the State, one other institution considered in the labour market is unionization. The role of radicalism in the NGO movement in the late 1970s and 1980s are the consequent organization and mobilization of rural labour which SSScannot be ignored in explaining the real wage rise as well as other changes like liberation of bonded labour and reduced proportion of attached labour.

Rural Development through Technological Innovations by NGO's

Creation of appropriate opportunities and assets for additional remunerative work and promotion of full employment in rural areas is a challenging task. Therefore there is a need for gradual up gradation of traditional economic activities[81] and their homogenous blending with modern technologies tools, techniques and trades. All these need application of science and technology, where professional scientists, technologists and planners have a vital role to play.

Rural based industries require a suitable mix of traditional and modern techniques of manual and mechanical operation of systemic research and development work and continued effort at improving the technologies involved. Here comes the role of NGO's who are involved in the village areas for technological development.

The application of technology involves :

- Selecting relevant technology from different alternatives;
- Making it adaptable;
- Providing science and technology support for its successful commercialization.

The rural economy can benefit quality from the introduction of technologies, which address the basic needs of people including income generation, food, shelter, clothing, which use local resources.

Following are various programs where the government involves NGOs for implementation and management of technology.

NGO's and its involvement in environment, agriculture and water supply :

"Krishak Bandhu Pump" was designed and developed by (International Development Enterprises) (IDE) a world wide enterprise development organization. The basic aim to improve the economic and social status of the common man as well as improve the environment to bring out improvement in the irrigation system. This organisation has been working in Orissa, U.P., Bihar, Assam and West Bengal since 1993 for marginalised farmers.

Volunteers in Technical Assistance (VITA) is a private international development organization. VITA places special emphasis on areas of agriculture and food processing renewable energy application, water supply and sanitation, housing and construction. They have developed simple technologies for water resources like getting ground water from wells, springs tube well, dug wells, water lifting and transport water storage, water treatment. Village Industry and Crafts are also developed.

Action for food production (AFPRO) is a NGO involved in promoting Biogas since 1904. This is in partnership between AFRO and Canadian Hunger Foundation. There are about 100 NGO's working in 14 states of India. Biogas technology is the apt option to meet the needs of the village. It is clean, convenient. It is used for cooking, lighting, and supply power for irrigation and small-scale industries. The effluent slurry is used as organic manure.

All India women conference (AIWC) is one of the oldest and pioneering women organisations working towards emancipation and empowerment of women working in rural areas. It conducts awareness and training programmes on biogas, solar cooking etc.

Advantages of NGOs

It is widely recognised that NGOs have several advantages.

More Actor Oriented : NGOs tend to take up activities which are needed for the people. NGOs are able to undertake need-based activities because they undertake studies relating to situation and needs of the people. They find out: Who are the poor? Why are they poor? What is to be done for alleviating poverty? The rigour with which the situational and needs assessment studies are carried out may vary across the NGOs. But, a starting point in the case of most of the grassroots NGO's is the articulation of the problems of the community in their project location. Most of the NGOs focus on the important problems that the poor and the marginalised community such as women, dalits and adivasis face in the locality.

Flexible in Methods and Practices : NGOs exhibit a high degree of flexibility in their functioning, method and practices because they tend to be local and small. The geographical area the NGOs tends to be small. A survey of 16 NGOs in Karnataka revealed the number of villages, four of them covered between 50 to 100 villages, and two of them covered more than 100 villages. Most of the NGOs work in one or two taluka in district. The available studies also suggest that this is a widespread phenomenon. Such a small geographical area enables the NGOs to be flexible. Methods and practices of formulation, implementation and monitoring of the programmes and style of working can easily be changed to suit the needs and aspirations of the community with which they are working, thus changing rural conditions. Being small, NGOs rarely face the problems of hierarchy and bureaucracy that restrict the government departments. NGOs respond swiftly and efficiently to local demands. Being locally based, they are aware of the local environment and are responsive to it.

Adopt Innovative and Participatory Approaches : One of the important contributing factors to the popularity of NGO's among donor agencies, government and development practitioners is that NGO development programmes tend to be innovative and emphasise on participation approaches. The innovative credit programme of the Grameen Bank at Bangladesh is implemented by all the countries including developed West (notably, USA and Canada). This programme is replicated on a large scale across the globe. Similarly, the Credit Management Programme of MYRADA influenced many NGOs to start similar programmes in their project areas. Their programme also influenced the NABARD to introduce bank-Self-Help Groups (SHG) linkage programme initially on a pilot basis, and now all over the country. NGOs (such as MYRADA), Centre for Appropriate Technology (CAT , Nagercoil, and Tamil Nadu) also developed innovative approaches in relation to watershed, appropriate technology, dry land development, etc. For instance, CAT has prepared a boat, which used much less energy than the conventional boats. Similarly, biogas was prepared by using the stems of bananas in the project area of RASTA in Waynad district of Kerala.

The strength of the NGOs lies in the fact that their structure and style of operation are such that the distance between leaders and members is minimal and democratic participative decision-making is rooted to the ground. NGOs are constantly designing innovative participatory approaches to elicit and enable the people to participate in the programmes. Several methodologies aimed at enabling the people's participation have been devised. The initial ideas relating to methods like Participatory Rural Appraisal (PRA), Participatory Learning and Action (PLA) were obtained by researches (notably Robert Chambers) from the project areas of NGOs. After some development of these ideas, NGOs again successfully adopted them and contributed to their further development. Because of these techniques and other NGO practices such as less hierarchy, flexibility, employment of local people, target group communities actively participate in the programmes and this contributed to the development activities being relevant to the people's needs and aspirations. The participation of the people also contributed to the reduction in costs.

More Focused in Development Work : NGO development tends to be more focussed as it is their principal goal. The goal confusion' is one of the most important contributing factors for the limited success of the government-initiated programmes in India. For instance, the main objective of Integrated Rural Development Programme (IRDP) was to provide income generating assets to the rural poor, and integrate the activities relating to raw material supply, skill provision and marketing.

But in practice, the IRDP could not integrate the activities. On the other hand, the NGOs normally have one principal goal, which could be poverty alleviation, mobilisation of the marginalised section to access government programmes, etc

Relative Independence : NGO development programmes enjoy independence; their governing boards are autonomous; and their development activities are primarily meant for the target group. The government programmes tend to be influenced by the local vested interests and political intervention. Under these programmes, the benefits meant for the poor are appropriated by the non-poor because the local power structures control these programmes. On the other hand, the methods and practices adopted in formulation and implementation of NGO development programmes, selection of the target groups, provision of services, etc., and relatively independent of the local power structures. This contributes to the target group participation and success of the programmes. The evidence on the autonomy of NGO programmes is, however, mixed; studies note that while NGO programmes enjoy relative independence as far as local power structure are concerned, they are influenced by donor and government policies.

Effective in Development Work : Because of the above, the NGO development programmes tend to be effective in reaching the poor, poverty alleviation and in cost reduction. The targeting is relatively good in NGO programmes. The studies reveal that 80 to 100 percent of the target group in GO project areas come from the poor categories as compared to only 60 to 70 percent in government programmes. The development NGOs have succeeded in :

- Breaking the isolation of the poor;
- Enhancing the productivity of assets and labour;
- Improving marketing opportunities for the produce such as milk, handicrafts, etc.; and
- Enhancing access to food, health, education and drinking water.

On the positive side, NGO development programmes have been most effective in the case of those poor with some initial endowment base. The NGOs are, however, not very successful in enabling the poorest among the poor with very little endowment base to begin with, to expand the means to overcome poverty.

Comparative Advantages : The evidence suggests that the NGOs have several advantages compared to the government. Unlike the government agencies, the NGOs seems to be highly motivated and are prepared to accept hardship as a challenge rather than as a punishment. With rather small size, selective tasks, personal leadership and flexible structures, voluntary agencies can innovate, adapt themselves to new circumstances experiment and face risks. A comparative study of economic development programmes for small farmers initiated by the government through IRDP programme and an NGO in Mangalore district, Karnataka concludes that the NGO programme achieved better results than the government through the proper identification of the problem, designing of the suitable and sustainable packages, proper appraisal and implementation of the project, meticulous follow-up, continuous training of the beneficiaries, relented coordination work at all levels by the project authorities. Another comparative study on self-employment programmes for rural youth initiated the government departments, NGOs, private organisations, and Grassroots Organisations in four districts in Karnataka reaches a similar conclusion that the NGOs, with commitment, dedication, missionary zeal, flexibility, etc., were more successful in implementing this programme than the other organisations.

Advocacy and Lobbying : NGOs play an important role in influencing the State policies by advocacy and lobbying through their networks. It is now recognised that a part of the requirement of a successful democracy is a strong civil society. Civil society can counterbalance the interests and actions of the State where it is necessary. Civil society is the arena in which the interests of different groups within the social formation can be presented through a wide variety of means and actions. NGOs are important organisations within civil society adding to its ability to influence and strengthen the process of development in Third World Countries. NGOs are therefore, important not just for the fact they can do' development better, but also because they can influence the perception, including that of the State, of what constitutes better development. The evidence shows that NGOs are playing a significant role in influencing the policies of the State at various levels, and counterbalancing the interests and actions of the state.

Fevord-K can be cited as an organisation, which has achieved significant results in checking the blind eucalyptisation' of all available land. The obsession with raising raw material for industry was such that even forest areas, which were unsuitable for raising eucalyptus, such as the Western Ghats, were planted with it on a massive scale. Farmers were also encouraged to grow eucalyptus in the name of local forestry. Vast stretches of lands, either government owned or common lands, which are normally used by villagers for grazing their cattle or as a source of other bio-mass, were also taken over for raising commercial species, especially eucalyptus. This meant alienation of the common lands from the locals as these developed' lands hardly met their needs like grass. A public sector company, Karnataka Pulpwood Ltd. (KPL), was formed for raising pulpwood for industry and was entrusted with about 16,000 hectares of government land for the locals over their common lands, and filed a legal suit against the government. Ultimately, after a protracted struggle over 5 to 6 years, the decision to wind up KPL was announced. In the process of this campaign, the Social Forestry wing of the Forest Department shifted its emphasis from farm forestry to community forestry and from commercial species to trees favoured by local people.

Weaknesses of NGO Movement

Not withstanding these potential advantages, NGOs and their development programmes face a number of weaknesses. Some important ones are discussed below :

Spatial Limitation : One of the important problems that NGOs face is their spatial limitation. Quite simply this is to say that NGO development projects remain little more than dots on a map. The territorial space under the NGO is defined by a number of factors; these include the obvious such as the amount of finance available, the physical characteristics of the project, the organisational size and structure of the NGO. Other important reason is that the poor located beyond the project area of NGOs are not provided with any development assistance. In addition, there is the spatial dimension determined by the definition of target group. Not all poor or marginalised groups will be covered within the territorial location.

Lack of Good Governance and Transparency : NGOs need to have good governance. However, in the absence of constant pressure from below, NGOs can assume the paternalistic role and a shift of priority from community to institution building. As the NGO expands the area and scope of activities, the leaders begin to dominate the NGO. The NGOs become semi-bureaucratic and hierarchical where initiative and decision-making gets confined to the leaders. This will have adverse impact not only on flexible functioning but also on development of new leadership. Often, the views and concerns of young staff members in the organisation are overlooked, or even suppressed. The young staff members either leave the organisation and/or remain as non-entities. Further, the staff do not the know budgets proposed and sanctioned, utilisation of budget, etc.

Patchwork-Quilt Phenomenon : A phenomenon of Patchwork-Quilt' is noticed and this has implications on the provision of services to all the needy in all the regions. It has been pointed out that the large, influential and well-funded NGOs may not be able to concentrate resources in regions and sectors that are most important for national development. If there is a concentration of NGOs in the already developed regions, the poor in the backward regions may be bypassed as far as service provision is concerned in the context of weak central overseeing. This is already happening in India. Although the economic and human development indicators are better in south India compared to central Indian states of Bihar, Madhya Pradesh, parts of Uttar Pradesh and western State of Rajasthan, the concentration of NGOs is more in the former compared to the latter. This is borne out by the fact that out of 12,136 organisations receiving foreign funding and reporting to the Home Ministry in 1996-97, there were 5,721 organisations in South India, while the number was only 1,779 in BIMARU region. As a result, service delivery is taking place in the already better off region while the poorer states are left out.

Inability to Reach the Poorest : Are NGO programmes more successful in reaching the poorest? A study of nine NGO savings and credit programmes spread across the country suggests that the ability of these programmes to provide credit to the poorest was limited. The emphasis on savings linked to credit, and enabled many of those poor with ownership to some productive assets to access more credit. A majority of the landless could not borrow because loans were given for the land-based activities of crop production, and off-farm activities and that they were discriminated by the group's'management committees.

Antagonisitic Attitude Towards the State : NGOs often seek to supplant the State as the provider of basic services and development programmes and thereby weaken the political relationship between people and their government. Further, there is a tendency of NGO social projects to produce a form of communalism because of their self-enclosed and self-valuing nature and the fact that the NGO involved begins to encompass the whole of the life of their members. In such instances, it is not merely that the members cease to look to the government for services, etc., they positively turn away from the government.

Palliative Nature of Service Provision : The evidence suggests that service provision by the NGOs tends to be palliative. A study on income generation programmes undertaken by NGOs across the country reveals that most of the NGOs could only facilitate the undertaking of subsistence activities, and income from such activities was either equal to or less than the existing wage income. These activities made difference to the people in so far as they could be undertaken during the lean season and that the problem of seasonal unemployment could be, to some extent, solved. The income from the economic activities promoted by NGOs formed only a small proportion of the total income of the member families. These findings suggest that the development programmes of NGOs can only be palliatives. Such an impact would be a barrier to the basic changes in the ownership of land and capital assets that are essential if significant economic and political changes are to occur.

Limited Ability to Influence Macro-Policies : Individual NGOs are rarely in a position to influence government policies at various levels. In addition, rarely do they seek to influence policy, because their existence and tolerance by the State is based upon the non-controversial, apolitical involvement in development. Furthermore, NGOs rarely cooperate and liaise with one another, tending to view each other as competitors for donor funding, State tolerance, and on occasion, local influence and space. Consequently, they fail to emerge as genuine macro actors at regional or at national levels, seeking to represent their members' interests in the political process of governance and more general development. In so far as they engage in politics, it is through the existing political parties by having members elected to local councils and mobilizing votes for party members at higher levels of government assembly. This is not enough as the problem lies in the institutions, in the formulation of policy and the formal and informal processes of political bargaining that surround it. Unless NGOs come together in networks, they will have little influence or impact on development or development policies at the more general level.

Lack of Accountability : Accountability should be equated with accountancy. The term accountability implies the extent to which NGO activities and programmes seek to fulfill the objectives with which NGOs were started. Accountability can be upwards (i.e., to the government, donors and governing board , sideways] to the interested public, media, etc.) and downwards (to the people, staff, etc.). it has been that NGO projects, while accountable to the donors, are hardly accountable to the people with whom they work. A study of 16 NGOs in Karnataka concluded that there is a mis-match between objectives and activities of these NGOs, thus suggesting that NGO projects are not accountable to the people. This implies that the NGOs projects do not emerge on the basis of situation and needs of the poor with whom they work but, influenced by donor priorities and policies.

Inappropriate Models : The models adopted and approaches followed by a vast majority of the NGOs make it difficult for the targeted beneficiaries community to absorb and sustain the activities when the funding stops or when NGO withdraws (or collapses/winds up). Instead of strengthening the autonomy and independence, one often finds the NGO models prepetuating dependency on NGOs, which, in turn, depend on external donors. This threatens the very fabric of self-reliance and sustainability.

Initiatives for NGOs

The involvement of NGOs and participation of people in the planning and execution of various anti-poverty and minimum needs programmes are very necessary for rural development. If these NGOs are provided managerial and professional skills and adequate funds are provided to NGOs they can play pivotal role as an intermediary development agency in the sphere of rural development. Following are some of the initiatives for the NGOs, which can make them more efficient in planning and implementing the rural development programs.

Democratic Decentralization and Empowerment

- Organising and activating women's participation in development programmes.
- Promoting backup support to the weaker sections of the society.
- Acting as facilitators to supplement activities of Panchayats and other decentralized institutions of people.
- Helping and advocating the cause of Panchayat Raj Institutions, women groups and youth collectivities.
- Act as pressure groups for grassroots sustainable rural development based on equity.
- Help Panchayats in identification of the real beneficiaries under target schemes.
- Act as bridges between Panchayats and government agencies by knocking out red tapism and officialdom.
- Provide backup support to the beneficiaries under the target methodology.
- Minimize areas of conflict between government agencies, PRI's and NGO's leading to smooth development work.

Empowerment of Women

- Confidence-building could strengthen and spread awareness among village women about existing avenues of socio-economic development programs through the method of person to person communication.
- The cooperation of literate and enthusiastic village women could be enlisted to accelerate government empowerment programs through the help of Village anchayats and self help groups so as to achieve optimum results.
- Such empowerment programs could focus on priority areas e.g.,
- Education and literacy;
- Vvocational training;

- Eentrepreneurship, production and marketing;
- Health, hygiene and sanitation;
- Leadership building;
- Micro credit financing;
- Appropriate village technologies.
- Capital building activities could be based on voluntary contributions, as far as possible, in addition to the help by government agencies, banks etc.

Appropriate Technology and Rural Development

- Identifying peoples' problems and designing appropriate technology adoption at village level. NGO's can organise target groups for particular areas with specific programmes. E.g. production and marketing, processing and value adding activities and water conservation, use of renewable energy resources etc.
- Thrust could be given to educational facilities from primary to higher levels where world of learning and world of work would coalesce.
- NGO representative could be trained in entrepreneurship development by reputed institutions.
- Before implementing development program NGO's should consult beneficiaries, representatives of Panchayats, target groups, neighbouring people etc. for income and employment generating activities.

Environmental Education and Health Care

- NGO's could identify environmental problems viz., water scarcity, water pollution and water conservation, air pollution, waste disposal etc. in working localities. This can help in preparing a priority scale of needs and problems.
- NGO's should try to evolve and adjust alternatives-traditional and technological methods.
- NGO's could play an important role in motivating people towards better preventive health practices and assist in making basic health care more and more accessible at the grassroots level.
- They can help improve general sanitation of a village. They can disseminate knowledge about nutrition, immunization, and disease prevention and reproductive health requirements.
- NGO's should function as catalysts and facilitators in augmenting community awareness level about environmental sanitation and to hand minor health minor problems and ailments with first aid etc., with emphasis on the idea that prevention is better than cure.

Conclusion

Despite the technological advancement taking place in agriculture, rural poverty tends to get worse. The poverty alleviation programmes seem to have been concieved, designed and implemented with the view to provide immediate and short term relief to poverty groups who have not benefitted by the trickle down effects of growth. Anti-poverty programmes in rural areas have multiplied but the overall effect on target population have been negligible.

Rural Development Programmes in developing countries have not been rooted to the structural problems to become effective enough to yield the desired results. A wide range of investigations into the incidence of poverty unemployment offer a shocking results on the performance of the various programmes in that, there has been no significant reduction in rural poverty, unemployment and inequities.

An alternative perspective based on structural change in agriculture with appropriate change in rural institution, participative development in the grassroots level, through the diffusion of technology and knowhow. Possibility of change and modification of the government mechanism for the successful management of rural development should be explored.

Because of the above, NGOs have come to occupy a central position in facilitating development at the local level, and hence, have considerable space to initiate development directed at improving the condition of the more marginalised and disadvantaged social groups. This is one of the important opportunities that the NGOs have.

References

S.K. Chatterjee : Concepts of Development and Modernisation : Development Administration, Chap. 4, p. 34-35.

Fred W. Riggs, The Idea of Development Administration,' in Edward W. Weidner, Development Administration in Asia, Durham, N.C. Duke University Press, 1970, p. 25.

Fred W. Riggs, Further Considerations on Development', Administrative change Vol 4 No. 1 July-December 1976. p. 2.

Ibid, P. 3.

V. Sivalinga Prasad, K. Murali Manohar : Fred W. Riggs : Administrative Thinkers.

S.K. Chatterjee, op.cit, p. 86-37.

Shriram Maheshwari, Rural Development in India : A Public Policy Approach New Delhi : Sage Publications, 1985 , p. 13.

Ibid.

Vasant Desai, Rural Development; Issues and Problems, Vol I, p. 381.

Hoshiar Singh : Rural Development, Concept, Strategy and Approaches : Administration of Rural Development in India (ed), Sterling Publishers, p. 3.

Rural Development Programmes in India, A Paper of Government of India, Ministry of Agriculture and Irrigation] Department of Rural Development, New Delhi, 1978 , pp. 1-2.

World Bank, Rural Development', Sector Policy Paper, May 1975, p. 3.

R.T. Tiwari and R.C. Sinha, Rural Development in India, Ashish Publishing House, New Delhi, 1988, p. 1.

Uma Lata, She Design of Rural Development - Lesson from America, Baltimore : John Hopkins University Press, 1975, p. 20.

James H. Coops, Rural Sociology and Rural Development', Rural Sociology, Vol 37, No. 4, Dec. 1972, p. 515.

Robert Chambers, Rural Development : Putting the Last First, London Longman, 1983, p. 147.

See R. Hooja, The District as a Planning Unit, Style and Locus' and N.R. Inamdar, District Planning in Maharashtra' both in Indian Journal of Public Administration, XIX] July-Sept. 1974, P. 393-406, 320-27, Sudipo Mundle, District Planning in India, Delhi, Chapter II.

See Growth Centres and Rural Urban Continuation A.D. Moodie] (ed) Approaches to Rural Development, Bombay 1976, p. 49.

Bruce F. Johnson and Peter Kilby, Agricultural Strategies : Rural-Urban Interactions and the Expansion of Income Opportunity, Paris, OECN, 1973, p. 15.

Inayatullah, Approaches to Rural Development - Some Asian Experiences; Kuala Lumpur, Asian Centre of Development Administration, 1979.

Hoshiar Singh, op.cit, Rural Development in India, An Historic View, p. 13.

B.B. Mishra : District Administration and Rural Development, Delhi, Oxford University Press, 1983, p. 6.

Ibid, P. 387.

B. Rambhai : The Silent Revolution, Delhi, Jiwan Prakashan, 1959, p. 10.

Evaluation of Community Development Programme in India.

Prabhakar Singh, Community Development Programme in India, New Delhi, Deep and Deep Publication, 1982, p. 34.

B. Rambhai, op.cit, p. 14.

Pyarelal, Mahatma Gandhi, The Last Phase, Vol II, Ahmedabad, Navjiwan Publishing House, 1963, p. 551-52.

B. Rambhai, op.cit, p. 15.

S.C. Jain, Community Development and Panchayati Raj in India, New Delhi, Allied Publishers, 1967, P. 56.

A. Major, Pilot Project in India, University of California Press, 1958, P. 37.

V.T. Krishnamachari, Community Development in India', New Delhi, Government of India, Publication Division, 1958, p. 12.

S.K. Singh : Strategies for Rural Development : An Overview, p. 72-73.

Durgesh Nandini : Rural Development Administration (ed), Conceptual Framework of the Study, Rawat Publications, Jaipur, 1992, p. 10-11.

S.N. Mishra, New Horizons in Rural Development Administration Delhi, Mittal Publications, 1989, p. 2.

Government of India, Planning Commission, Third Five Year Plan Document, p. 316.

Kishore Chandra Padhy, Rural Development in Modern India New Delhi, B.R. Publishing Corporation, 1986 p. 89-96.

Government of India, Planning Commission, The Fourth Five Year Plan, p. 175.

Nadeem Mohsin, Rural Development Through Government Programme, Delhi, Mittal Publications, 1985, p. 15-16.

Government of India, Minority of Agriculture, Department of Rural Development, Annual Report 1988-89, p. 2.

Vasant Desai, Rural Development; Programmes and Strategies, Vol II, pp. 88-89.

Vasant Desai, Rural Development, Rural Development Through the Plans, Vol V, p. 47.

Kishore Chandra Padhy, Rural Development in Modern India, p. 124.

Ibdi, p. 246-50.

Ibid, p. 127.

Government of India, Ministry of Agriculture, Department of Rural Development, Annual Report 1988-89, p. 4.

Government of India, Ministry of Agriculture, Department of Rural Development, I.R.D.P.-1987, pp. 9-16.

IRDP Success Limited, The Hindustan Times, 12th June, 1989, New Delhi.

Government of India, Ministry of Agriculture, Department of Rural Development, A Manual on Integrated Rural Development and Allied Programmes of Training of Rural Youth for Self Employment (TRYSEM) and Development of Women and Children in Rural Areas (DWACRA), November 1989, p. 20.

Government of India, Ministry of Agriculture, Department of Rural Development, Annual Report 1988-89, p. 2.

Government of India, Ministry of Agriculture, Department of Rural Development, Manual on National Rural Employment Programme (N.R.E.P.) and Rural Landless Employment Guarantee Programme (R.L.E.G.P.), Oct. 1986, p. 1.

Ibid.

Government of India, Ministry of Agriculture, Department of Rural Development, Annual Report 1988-89. p. 44.

Anti-Poverty Programme at a Glance, Gramin Viaks Newsletter, Vol. 5 No. 2-3. Feb - March. 1989. p. 37.

Government of India, Ministry of Agriculture, Department of Rural Development, A Manual on Integrated Rural Development Programme and Allied Programme of Training of Rural Youth for Self-Employment (TRYSEM) and Development of Women and Children in Rural Areas (DWACRA) November 1988. p. 24.

Ibid.

Government of India, Ministry of Agriculture, Department of Rural Development, Annual Report 1988-89. p. 2.

Government of India - Ministry of Programme Implementation, The Twenty Point Programme 1986 Perpectives and Strategies, The Cutting Edge of the Plan for the poor, preface.

Government of India, Ministry Agriculture Department of Rural Development, Guidelines for Implementation of Accelerated Rural Water Supply Programme, August 1989. p. 1.

Anti-Poverty at a Glance', Gramin Vikas Newsletter, No. 5, No. 2-3, Feb - March 1989, p. 37.

Jawahar Rural Job Plan Launched', The Hindustan Times, New Delhi, April 29, 1989.

A Manual on Jawahar Rozgar Yojana, Department of Rural Development, Ministry of Agriculture, Government of India, New Delhi, August 1989, p. 2.

Ibid p. 3.

Gramin Vikas, Obligatory Programmes, Government of India, Rural Development Ministry, Produced by the Directorate of Advertising and Visual Publicity, Ministry of Information and Broadcasting for Ministry of Rural Development, Government of India and Printed at Brijbasi Art Press Ltd., New Delhi, No. 3/14/2000, July 2001, p. 1.

Ibid.

Hoshiar Singh : Rural Development Concept, Strategy and Approaches; Administration of Rural Development in India, Sterling Publishers, 1966. p.10.

D. Rajasekhar, Non-Governmental Organisation (NGO's in India. Opportunities and Challenges; Journal of Rural Development, Vol 19 2 , NIRD Hyderabad), pp. 249-250

Murthy R.K. and Nitya Rao, 1997 : Indian NGO's, Poverty Alleviation and their Capacity Enhancement in the 1990's : An Institutional and Social Relations Perspective, New Delhi, FES. Indian express.com, Aug, 2015

Bedi, Narinder, (1999), Development of Power, In D. Rajasekhar (ed) Decentralised Government and NGO's : Issues, Strategies and Way Forward, Delhi, Concept.

Fernandiz, Aloysius P. 1996 : The MYRADA Experiences : Working with the Government in Multilateral and Bilateral Proejcts, Bangalore-MYRADA.

Ibid.

M.K. Dubey, Rural and Urban Development in India : Centrally Sponsored Schemes] (ed) Common Wealth 2000, pp. 267-268.

Ibid p. 270-271.

Ibid p. 273-274.

Ibid p. 275-276.

Mahipal : Role of Voluntary Organisations in Rural Development, Social Change, September 1991, Vol 21, No. 3, p. 27.

Mahipal : 1990 Group Enterprenuership among IRDP Beneficiary, Gramin Vikash News Letter, Department of Rural Development, March 1990. pp. 10-11.

Oley S.K. (1985) Voluntary Action or Collusion ? Kurukshetra Vol 34, No. 1, Oct. 1985, p. 5.

In 1970 the Government of India Launched the Rural Works Programme] RWP in drought prone areas as an employment-oriented programme. However, the RWP was unable to provide employment avenues on regular basis. In 1972-73 the RWP was modified and given the name of drought prone area programme. It was originally designed and employment generation programme but letter shifted focus to development works. So that It may provide permanent solutions to the problems of drought rather than piecemeal solutions to the problem of unemployment.

WHO's Committment to improving the Conditions that are exclusive to Women such as Pregnancy and Childbirth is enshrined in its constitution The Safe Motherhood Initiative SMI launched in 1987, by WHO, UNICEF UNFDA, the World Bank and a host of Concerned Organisations, drew global attention to the hidden and tragic inequity of Maternal ill health.

Balkrishna, Lalitha and Mishra Meenu : 1998, Role of NGO's in Sustainable Rural Development : IREDA News, Vol 3, July-Sept 1998, p. 47.

Alka Srivastava : Non-Governmental Organisations and Rural Development Social Action, Vol 49. Jan-March 1999, p. 27-44.

Rajsekhar D. 1998 : Rural Development Strategies of NGO's, Journal of Social and Economic Development 1(2).

Farrington, John et al (ed) 1993, the Reluctant Partners ? Non-Governmental Organisations, the State and the Sustainable, Agriculture, London Reutledge.

Ibid.

Murthy R.K. and Nitya Rao, 1997, op.cit, p. 57.

Ibid.

Sangitha S.N. 1990 : Self Employment Programme for Rural Youth : The role of Non-Governmental Organisations NGO's IIMB Management Review, 5 (2) .

Webster, Neil, 1995 : The role of NGO's in Indian Rural Development : Some lessons from West Bengal and Karnataka, The European Journal of Development Research 7 (2).

Joshi S. et al] (ed) 1997, Experiences of Advocacy in Environment and Development, Banglore, DSI and NOVIB.

Ibid.

Robinson M. 1993, Governance, Democracy and Conditionality : NGO's and the New Policy Agenda in A. Clayton] (ed), Governance, Democracy and Conditionality: What role for NGO's ? Oxford, INTRAC.

Ibid.

Namrata, 1999 : Income and Employment Generation Programmes of NGO's in India, Banglore : Development Support Initiatives, Forth Coming.

Riddel R. and Mark Robinson, 1992 : The Impact of NGO Poverty Alleviation Projects : Results of Case Study Evaluations, Working Paper No. 68, London, Overseas Development Institute.

T.N. Dhar, NGO's : Their Roles, Responsibilities and Problems; Quarterly Newsletter, Dynamic Administration (ed) S.P. Gupta, U.P. Regional Branch : Indian Institute of Public Administration (Jan-March 2000) , 49th Issue.

CHAPTER-3

Some Major Voluntary Agencies in Marathwada

Introduction

In this chapter the meaning and need of the NGOs is studied illustrating the areas of NGO's involvement in rural sector. The chaper enlists some of the voluntary agencies in Maharashtra and Marathwada. The development perspective in Marathwada is also briefed. The voluntary agencies vistied and observed by the researcher are given in detail stating their organisation, approach, area of operation, objectives and activities.

Voluntary organisations have played a vital role in social progress and they have a long history of active involvement in the promotion of human welfare and well being.

Need of Voluntary Agencies

Rural development programs initiated by the government since independence have not reduced poverty substantially. One of the important contributing factors for the failure of rural development programs were the absence of involvement of the people for whom the programs were meant. The need for micro level institutional arrangements, to involve people in formulation, implementation and monitoring of the programs is therefore stressed in several quarters.

A sizeable number of NGOs have involved in welfare and development work in rural areas. Development of village industries, agriculture and allied activities through technological innovations are the common activities of large number of agencies engaged in rural development. Different approaches were adopted to achieve rural development in India. The government carried out integrated plans for rural development and at the same time the voluntary agencies also initiated rural development activities in specific localities. NGOs by virtue of being small scale, flexible, innovative and participatory are more successful in reaching the poor and in helping in alleviation of poverty.

Meaning

The word Non- Government Organisations (NGOs) is used for those agencies functioning outside the government sector. For voluntary action we need organisations which are indigenious, decentralised and non-statutary with direct participation of the beneficiaries or target groups in the programme. Nowadays voluntary action is concieved in a broader and combined connotation i.e." Admixture of voluntarism as well as activism ".

Scope

The reach of voluntary action in rural society is now expanding with the observed needs and scope. NGOs are credible in the sense that they motivate rural masses for a change of attitudes and perception, secondly they generate awareness about the rural development programmes on one hand and alternative paths on the other. Thirdly they indulge in individual and community actions to transform the plans into reality. Lastly, through repeated reinforcement they induce refinement over a period of time.

The Areas of NGO's Involvement

Studies on actual working of some of the voluntary agencies in different parts of India reveal their modus operandi and their impact on contemporary rural society. An observation of studies reveals that voluntary agencies are engaged in :

- **Agricultural change**

  Through breaking down traditional barrier, adoption of new technology use of fertilizers and seeds. Soil conservation storage, marketing and supply of agricultural inputs.

- **Development of Animal Husbandry**

  Through dairy development, poultry, fisheries, piggeries etc.

- **Development of Village Industries**

  Through promotion of village and khadi industries, promotion of forest based industries, extension services, vocational training etc.

- **Economic Development Programmes**

  Through bhoodan and sarvodaya , rural employment, promotion of cooperatives, integrated rural development.

- **Education**

  Through pre-school education, school and college education, adult education, community education, library service etc.

- **Health and Medical Services**

Through maternity and child health, dispensaries, hospital, leprosy central and rehabilitation, environmental hygiene, community health services etc.

- **Social Services and Amenities**

Through housing construction of streets, drinking water supply, construction of community centres, electrification bio-gas etc.

- **General Welfare Services**

Like welfare of women] children, aged welfare, unemployed aged youth scheduled castes and scheduled tribes etc.

- **Family and Child Welfare**

Through nutrition programmes, training of bal sevikas/gram sevikas etc.

- **Institutional Services**

Environmental protection through preservation of ecosystem, chipko movement, social forestry, control of pollution etc.

- **Other services**

Eradication of social evil prohibition, legal assistance, settlement of disputes, promotion of people's institutions, social action, cultural activities, research, research publications, and other promotional activities.

It is also studied that, the fifty five years of our development efforts have shown us that people cannot remain passive recipients of government aid and poverty alleviation programmes. A proactive effort and organisation of disparate groups/individuals, especially the disadvantaged sections of society through the efforts of NGOs would help overcome the curse of poverty, ignorance and lack of awareness in our society. NGOs have the advantages of flexibility of operations, limited expenditure on staff, emphasis on quality over quantity even while being focussed on achieving targets and the ability to attract funds from diverse sources.

In India there are innumerable voluntary agencies working in various fields for the development of the people and the country. There are number of registered and unregistered, small scale and large sector agencies in India. There are also networks of voluntary agencies grouped as network of rural development agencies, and agencies in watershed etc.

AVARD is the Association of Voluntary Agencies of Rural Development. It is an open secular, non-political, non-party all India Federation or Network of about 550 autonomous voluntary agencies engaged in rural development across the country. Some of the agencies working for rural development in Maharashtra are:

- Academy of developmental sciences
- Acil - Navsarjan Rural Development Foundation.
- Afarm
- Agro Industries and Tribal Welfare Foundation.

- Centre for Technological Alternatives for Rural Areas.
- Gramyan
- Green Future Foundation.
- Karzat Taluka Govanshi Sudhar Mandal.
- Maharahstra Arogya Mandal.
- Mahatma Phule Agriculture University.
- Nimbkar Agriculture Research Institute.
- Pravara Sahakari Sakhar Karkhana Ltd.
- Pryog Parivar
- Society for Action in creative Education and Development.
- Spinner Plantation Pvt. Ltd.
- Verala Irrigation Development Project Society.
- Apeksha Homeo Society.
- Social Centre.
- In Maharashtra there are also many voluntary agencies working in various fields of development like agriculture, integrated rural development, education, village industries, irrigation, livestock department, environment, family welfare and health, energy development, afforestation, science, technology and labour, etc. Following are some examples of some voluntary agencies working in the fields as follows :
- Bhartiya Agro Industries Foundation, Pune.
- Academy of Development Science, Raigarh.
- Centre of Science for Villages, Wardha.
- Indian Institute of Education] Castford and Vigyan Ashram , Pune.
- Jnana Prabhodhini, Pune.
- Pravra Institute of Research and Education, Ahmednagar.
- Rural Communes, Mumbai.
- Action for Agricultural Renewal in Maharashtra, Pune.
- Decospin Charitable Trust, Kolhapur.
- Gokul Prakalp Pratishthan, Ratnagiri.
- Gramodaya Sangh, Chandrapur.
- Institute for Rural Development and Social Services, Jalgaon.
- Mahatma Gandhi Shikshan Sanstha, Buldhana.
- Nimbkar Agricultural Research Institute, Satara.
- Sanjay Education Society, Dhule.

- Sarvodaya Samiti, Bhandara.
- Vidarbha Maharogi Seva Mandal, Amravati.
- Janiv Jagriti Sanstha, Satara.
- Karmaveer Shikshan Sanstha, Nasik.
- Shri Shivaji Shikshan Prasarak Mandal, Solapur.
- Verala Dairy Project Society, Sangli.
- Academy of Young Scientist, Nagpur.
- Banwasi Kalyan Ashram, Mumbai.
- Gram Vikas Seva Mandal, Chandrapur.
- Jai Malhar Krishi Vikas Pratishthan, Pune.
- Jalgaon District Child Welfare Council, Jalgaon.
- Jeevan Sansthan, Pune.
- Maitree Mandir, Ratnagiri.
- Prerana Pratishthan, Satara.
- Samajik Sadbhav Sangh, Vardha.
- Surya Community Centre, Thane.
- World Vision of India, Pune.
- Amhi Amchya Arogyasathi, Gadchiroli.
- Akhil Bharatiya Madhyam Vargiya Samaj Prabhodhan Sansthan, Thane.
- Adivasi Seva Mandal, Jalgaon.
- Bharatiya Adim Jaati Sevak Sangh, Nagpur.
- Disha Kendra, Raigarh.
- Foundation for Agro Ecologic Science, Buldhana.
- Gram Vikas Sanstha, Bhandara.
- Gramin Vikas Shikshan Sanstha, Amravati.
- Gramin Yuvak Vikas Prasarak Mandal, Bhandara.
- Indian Institute of Youth Welfare, Nagpur.
- Institute of Regional Development Planning, Amravati.
- Institute of Rural Reconstruction, Thane.
- International Buddha Education Institute, Nagpur.
- Jeejamata Mahila Mandal, Ahmednagar.
- Jungle Kamgaar Society, Adiwasi Aushadhupchar, Nasik.
- Karmaveer Shikshan Sanstha, Mumbai.
- Krantiveer Chaphekar Smarak Samiti, Pune.

- Krishak Seva Sangh, Ahmednagar.
- Krishi Seva Kendra, Yavatmaal.
- Mahatma Phule Samaj Seva Mandal, Solapur.
- Paramparagat Aushadhi Sanshodhan Vikas Kendra, Gadchiroli.
- Pragati Pratishthan, Thane.
- Sategaon Education Society, Amravati.
- Satpuda Vikas Mandal, Jalgaon.
- Appropriate Rural Technology Institute, Pune.
- Amarshakti Gram Vikas Shikshan Sanstha, Nagpur.
- Bhagwati Gramin Mahila Sangh, Ahmednagar.
- Bharatiya Gramin Vikas Mandal, Solapur.
- Following are the voluntary agencies working in Marathwada's Rural Development.
- Ashish Gram Rachana Trust, Aurangabad.
- Children and Women Welfare Association of India, Osmanabad.
- Nari Prabhodhan Manch, Latur.
- Prabhat Shikshan Prasarak Mandal, Nanded.
- Rashtriya Jagriti Bahuddeshiya Seva Sanstha, Latur.
- Rural and Urban Slum Social Service Implementing Association, Latur.
- Socio-Economic Development Trust, Parbhani.
- Vijay Vachanalaya Trust, Osmanabad.
- Public Progressive Development Circle, Nanded.
- Sanskriti Samvardhan Mandal, Nanded.
- Gandhi Niketan Shikshan Sansthan, Nanded.
- Padamshree Youth Service Institution, Latur.
- Progressive Friend Circle, Nanded.
- Sahayog Samwadini Rehabilitation Centre, Osmanabad.
- Sarwangin Vikas Sanstha, Latur.
- Society for Action in Creative Education and Development, Aurangabad.
- Antarbharati, Latur.
- Chetana Jan Vikas Mandal, Latur.
- Jai Mata Sudhar Mandal, Osmanabad.
- Lok Vikas Mandal, Osmanabad.
- Mahila Samasya Niwaran Mandal, Nanded.

- Matashree Gangadevi Mahila Samaj Seva Shikshan Sanstha, Latur.
- People Institute of Rural Development, Latur.
- Samaj Bharati, Latur.
- Samayak Knayan Prasarak Sanstha, Beed.
- Social Health Youth Committee, Latur.
- Bal Vikas Mahila Mandal, Latur.
- Bhartiya Gramin Punarrachana Sanstha, Aurangabad.
- Blue Bird's Education Society, Latur.
- Gramin Vikas Mandal, Nanded.
- Gramin Vikas Shiksha Sanstha, Latur.
- Indira Yuva Mandal, Aurangabad.
- Jan Jagrat Mandal, Hipalnari, Nanded.
- Jan Seva Sanstha, Aurangabad.
- Jan Vikas Mandal, Nanded.
- Mahatma Phule Social Education and Research Institute, Beed.
- Manav Vikas Mandal, Osmanabad.
- Manav Vikas Prakalp, Parbhani.
- Marathwada Gramin Vikas Sanstha, Latur.
- Rajiv Gramin Vikas Mandal, Nanded.
- Rural Development Centre, Beed.
- Rural Development Circle, Nanded.
- Sahyadri Shikshan Prasarak Mandal, Nanded.
- Sandhi Niketan Shikshan Sanstha, Nanded.
- Sarwangin Vikas Mandal, Nanded.
- Savitribai Phule Seva Kendra, Latur.
- Sheti Ani Gramin Vikas Sanshodhan Mandal, Osmanabad.
- Shri Ganesh Shikshan Prasarak Mandal, Latur.
- Shri Vitthal Shikshan Prasarak Mandal, Latur.
- Shri Yoganand Shikshan Prasarak Mandal, Jalna.
- Smt. Narsabai Mahila Mandal, Nanded.
- Society for Education in Values and Action] SEVA , Aurangabad.
- Surya, Nanded.
- Vidhayak Kendra, Latur.

Development perspective of Marathwada

In 1960-61 the Bombay State was split up and Maharashtra State was formed. Before, that development ideas were being formulated at the national level and a few development blocks were established in each district on a pilot basis. The standard pattern of development blocks was soon extended to all the Talukas of Marathwada and several new welfare programmes were introduced for rural areas. The then Chief Minister of Maharashtra Y. B. Chavan, declared, afresh, the State government's commitment to the Nagpur agreement and announced several programmes to remove the backwardness of Marathwada and Vidarbha. Marathwada remained backward, because Nizam's had deliberately neglected it because Maratha's, were Nizam's enemies .

Geographically the region is situated between 17^0 35N and 20^0 40 N latitudes and 74^0 40E and 78^0 15E longitutudes. Marathwada forms the central portion of Maharashtra. The area of Marathwada has a total of 64,717.9 Sq Km out of which 63,309.3 Sq Km is rural area. Marathwada is one of the five regions of Maharashtra which comprises seven districts, viz., Aurangabad, Jalna, Parphani, Nanded, Beed, Latur and Osmanabad. The entire region formed the part of the princely State of Nizam, then known as Hyderabad State, which was merged into the Indian Union in 1948. The origin of marathi culture is found in the history of Marathwada. Western Maharashtra dominated the State of Maharashtra economically and politically due to Shivaji. This region was a land of saints, a nursery of culture and seat of empire before Shivaji. The world renowned caves of Ajanta and Ellora are found here.

According to Dr. Kaldate, Marathwada is backward among the backward areas of Maharashtra because it is lagging behind in four major directions - population control, literacy, irrigation and social and educational services. These inadequacies have gone on increasing because of the lack of public participation and lack of mobility on the part of social and political leaders.

According to Dr. R. P. Kurulka agricultural sector is the most dominant sector in the economy of Marathawada region of Maharashtra State. In 1978-79, about 60% of the regional income originated in the primary sector as against the 32% at the State level. In 1991-92, the share of primary sector in Marathawada has come down to 34% while at the State level, it has come down to 21% only. This clearly indicates a changing structure of the Marathawadas' economy.

As the irrigation system of Marathawada is concerned, till 1956, surface irrigation in Marathawada was almost negligible. But, since the formation of the State of Maharashtra, a number of small, medium and large irrigation projects have been completed in different parts of the region.

In the recent developments, government of Maharashtra has formulated number of policies and programmes for the upliftment and empowerment of rural sectors. For example; special programmes: 42-Point Programme (1980), Committee on Regional imbalance in Maharashtra State (1984) , Creation of the Statutory Regional Development Boards (1994), Appointment of "The Indicators and Backlog Committee" (1994), Reconstitution of The Indicators and Backlog Committee (1995), New Economic Policy (1991), New Industrial Policy (1993).

Problems of Marathwada Region

In the villages of Marathwada it has been estimated that 62 per cent of the land owning class possess only 19 per cent of the land while 25 per cent own more than 70 percent . one of the reasons for the government allotting43 percent of public investment in the sixth plan to agricultural and rural development is to reduce such anomalies and place more credit at the disposal of small farmers who have hitherto been exploited by different agencies .

The problems of unemployment and poverty in India have been originated in rural areas. The Bhagwati Committee (1973) has rightly pointed out that, out of the total unemployed people, 80 percent come from rural sector. Dandekar (1973) in a conservative estimate indicated the unemployment figure as 19 million. A higher industrial wage rate encouraged migration from villages to the cities and newly developed towns.

Agricutural development, problem of unemployment and poverty are more complex in drought affected areas due to the uncertain elements in agriculture. The actual problems of the marginal farmers are as follows:

- Lack of awareness and understanding of dry farming technology and practices of cultivation
- Lack of financial resources
- Inadequate infrastructure
- Lack of technical advice within easy access
- Feeling of isolation and powerlessness
- Poor natural resources
- Scale of operation too small
- Lack of adaptive research
- Lack of markets
- Inacessibility of credit
- Lack of capital and thereby lack of enterprenurship
- Non-availibility of in puts
- Lack of employment opportunities in non-agricultural industries
- Unemployment during off season and under employment during the season
- Low wage rate
- Poverty
- Mal nutrition, health hazards, poor living standards, lack of housing, lack of recreation, lack of opportunities, low social and economic status, lack of capital investment for self emploment, use of traditional occupational skills, that is not very renumerative because of the competition with the industrial products.

In the empowerment of Marathawada Voluntary Agencies have a wide role to play. Their role in rural development is very important. It supplements the functions of Government of Maharashtra, where the government cannot reach out for various reasons, they undertake a number of projects. We can see them everywhere, in every issue which affects the lives of humans and the environment-children, women, dams, health, irrigation, watershed, education, old-age homes, orphanages, natural resource management, employment and income general and more efficient technologies and connected matters like empowerment and social justice.

Observations of Some Voluntary Agencies in Marathwada

After visiting some major NGOs in Marathawada region like: Marathawada Sheti Sahaya Mandal (MSSM), (Aurangabad), Manavlok (Ambejogai), Abhinav Vikas Sanstha (Fardapur), Grasp (Aurangabad), and Jannarth (Aurangabad), Integrated Health Management Pachod (IHMP,Aurangabad), Dilasa (Aurangabad) it was found that all these NGOs are doing tremendously well in their own region.

The basic instinct of their service motive and upliftment of poor in a sustainable form can be seen in these NGO's stated above. Each member of the NGO is drawn into the spirit of fulfilling the aims and objectives of the same.

They have a strong positive approach in the follow-up of every project undertaken. Among the NGO's stated above, many began with changing the orthodox opinion of the people and building awareness among the people about development, where they have done far better jobs compared to the government agencies.

Manavlok

"Manavlok" is located in the Ambejogai tahsil of Beed District, in Marathwada area. The nearest city is Aurangabad, 235 K.M. to the Northwest. The state capital Mumbai (Bombay) is about 500 K.M. to the west.

In the early 1970s, a socialist group of youth began seeking solutions to local, social and political problems with the general aim of achieving a more just and happy society. After nearly 20 years of effort and agitation's, however, they conceded that socio-political change was not possible without economic change. With this in mind, they decided and established in 1982 "Manavlok", a Voluntary organisation for socio-economic upliftment of rural-poor. A new experiment in rural development was launched. This philosophy is reflected in its name: "Manavlok" is the shortened form of "MArathwada NAVnirman LOKayat", where Navnirman' means "New creation" and Lokayat' denotes "People's opinion". "Manavlok" thus symbolises a new idea or an experiment in rural development, whereby villagers themselves locus of action and control[7].

Objectives

- To improve the socio-economic status of the rural people of the region.
- To empower women economically, socially and politically.
- To improve health conditions in the region.

- To improve the standard of formal and non-formal education and increase the academic standing of students.
- To build democratically working transparent organisations of rural poor, women and youth to solve their own problems locally.

Area of Operation

"Manavlok" works in over 180 villages in Ambejogai, Majalgaon, Parli, Keij, and Dharur Tahsils, focusing on marginal farmers, landless labourers and women. In 1993 it began work in the earthquake-affected areas of the Latur and Osmanabad districts. Few of these villages enjoy the benefits of the government irrigation networks. Rainfall in the target area is erratic and irregular, averaging 66.9 cm per annum. Temperatures range from 13 to 43 degrees centigrade. Sixty percent of the villages are situated in a hilly belt with red, stony alluvial soil, while the remaining forty percent are on undulating plains with medium to deep, black cotton soil. The population density is approximately 170 persons per square kilometre.

Socio-economic survey conducted by "Manavlok" in 1987 on a sample of 2500 families in the target area indicates that 28% of the population belongs to the dominant Maratha caste. The survey also revealed that 57% of the families had a gross annual income of less than Rs.4800/-. About 42% of the families are estimated to like on marginal farm incomes who together with another 20% who are landless migrate in lean periods to seek employment. Manavlok works with the people and for the people of the Marathwada region of Maharashtra. It aims to solve their socio-economic problems and achieve a better life for them.

Organisation and Approach

Manavlok is administered by a general body, an executive committee, and a full-time secretary. The general body meets annually and elects an executive committee triennially. This committee, in turn, meets quarterly and makes policy decisions in collaboration with the secretary. The secretary co-ordinates all programs, with the involvement of staff, and the community. The Honorary Executive Assistant implements the Manaswini' programs, which are special activities for rural women.

The approved programs are implemented through the seven Sub-centres: Salegaon in Osmanabad District, Pardhewadi in Latur District, and Patoda, Poos, Pisegaon, Yelda, and Ambejogai in Beed District. Each is headed by a sub-centre in charge. For efficiency, these centres are divided further into an administrative sector, a monitoring sector, a technical sector, an account sector, an education sector, a health sector, and a community organisation sector, each with a team of employees and a Sector In-charge'. For each project a separate ledger is maintained for accounts, all of which are computerised. A monitoring team observes activities and performs social audits. After completion of a program, an internal evaluation is conducted.

- Sector in-charge (SI)
- Area Community Organiser (ACO)
- Community Organiser (CO)
- Village Level Workers] Tai', Dai', sports teachers

Village level workers remain in the villages and organise activities as per decisions of Krushak Panchayats. Community organisers are responsible for overseeing the activities and budgets of five villages and meeting weekly with the A.C.O. for reporting and evaluation. Staff meetings are conducted monthly, in which reporting is done, problems are discussed, and plans are drafted. Nucleus of the system is the KP', the village level organisation of farmers and landless organised by Manavlok in each village.

The rationale is that KPs through democratic working build unity among villagers, provide a forum for their concerns, ensure the wise direction of funds, and provide a semblance of bottom-up management. In other words, the KP is

the springboard for total village development. Women are also members. Rallies and workshops are conducted regularly for KP members. Their participation is requested in the planning and budgeting of all programs. They, in turn, are expected to select according to need and criteria the beneficiaries of these programs, to annually elect a president and secretary, and to manage the village level activities with the co-operation of Manavlok Cs. Manavlok's activities are need-based, and planned and implemented only after discussion with the target group. No program is implemented without a community's or beneficiary's approval and participation.

Activities

Manavlok would develop its activities over an increasing scale, never straying from the tenet that programs be based on the specific needs of the people. Soil and water conservation, a mobile health clinic, community wells for marginal farmers, legal support for destitute and deserted women, skills training for women and landless, women and youth to solve their problems with available resources are just some of the important activities initiated. Since the rural economy is entirely dependent on agriculture, developments in agriculture will form the basis of any improvement in the plight of the common villager, Manavlok has thus chosen to focus its activities in this domain.

An integrated approach has been taken, as illustrated by our Employment for Community Development (EFCD) program. Wages are offered to villagers for digging new wells, desalting old wells, building check-dams, and generally developing the surrounding watersheds such that maximal monsoon precipitation is retained. This has several effects. The first, and most obvious, is in-creased employment and income for villagers. This is especially important for reducing their debilitating migration in lean periods. The second is an increase in farm income via increased irrigation capacity and crop productivity.

Another important activity is to assist small farmers, landless and especially women in generating supplementary income. Interest free advance is made available for short duration to establish small business or to provide skilled services. Training and even financial assistance is given for training in skill developments in sanitation, health, education, and general quality of life that can be made as a result of these economic gains. Finally, the attainment of some degree of income stability and self-determination serves as a sound psychological platform for the building of ideas related to environmental awareness, family planning, social rights and responsibilities, etc.

Besides the Krushak Panchayats' referred earlier, the rural women especially the poor are organised to form the "Bhumikanyamandals". "Bhumikanya" means daughters of mother land'. The members come together, discuss, voice and solve their problems and join the KPs in the Rural Development Programs. Similarly the rural youth are organised to form "Tarun Mandal". Tarun Mandal' means youth club.

Programs have also been initiated for the people of 31 villages devastated by the September 30, 1993 earthquake. Initial relief work included the provision of diagnostic medical camps, temporary shelters, basic tools and utensils, drinking water, emergency food supplies and sanitary measures. Recently an awareness program on water, sanitation and environmental care has been initiated in 46 villages under the guidance of Maharashtra Rural Water Supply and the World Bank's Environmental Sanitation Project.The present focus on long term rehabilitation includes support for seed, fertilizer, agricultural implements, education, health, psychiatric treatment, income generation programs (e.g. EFCD program), reconstruction of houses, and the formation of farmers' and women's groups. Other specific activities include the construction of a student hostel, a community hall, and earthquake resistant homes as shown in Table 2 and 3.

Manavlok : Table - 2: Summary Of Activities (1982-1997)

Sr.No.	*Activities*	*No. of Villages*
1.	Agricultural Development	
	Seeds and Fertilizer Distribution	185
	Afforestation	20
	Plant Distribution to Farmers	34
	Plantation aon Community Land	6
	Experimental Farming	3
	Support for Sprinklers	10
	Nydep and Composit Pits	35
	Support for Organic Farming	20
2	. Soil and Water Conservation	
	New Community Wells Dug	99
	Check Dams and Farm Ponds	49
	Overflows	41
	Mini Watershed Development	81
	Land under Watershed Program	81
3.	Income Generating Program	
	Support to Supplementary Income Scheme	75
4.	Skills Training	
	Wool and Carpet Weaving	30

Sr.No.	Activities	No. of Villages
	Masonary	4
	Bag Making	3
	Tailoring, Knitting and Embroidery	85
	MCED Training	18
5.	Women's Empowerment Program	
	Women Krushak Panchayat	15
	DWACRA Groups	12

Source : Report of Manavlok modified.

Cattle camps : Table - 3

Sr.No.	Year	Cattle Camp	Cattle per Day	Beneficiary Farmers
1.	1985	2	1500	750
2.	1986-87	4	5000	1000
3.	1992	13	4472	1321

Source : Report of Manavlok modified.

Support organisations

The bulk of operating funds are received from non-government donor agencies, including OXFAM, TDH, ACTION AID, NOVIB, HELPAGE, PLAN INTERNATIONAL etc. Some support is received from semi-government agencies like CAPART Delhi, and PAD Maharashtra. Also, the Zilla Parishad, Beed has provided support for the construction of bio-gas plants and the World Bank for the construction of sanitary blocks. Funds are also raised from local communities.

Nirman

The platform of Nirman was established in 1994 for co-ordinations of community based participatory integrated development for managing their own resources. The indigenous knowledge base, self-initiated development efforts supported with out experience have borne fruits of "Integrated Sustainable Development".

Objectives

Integrated sustainable development is the prime objectives of Nirman. The objectives of the organisation are:

- Health for all
- Essential education and literacy
- Land capability based agricultural development
- Conservation and management of natural resource
- Livelihood development
- Gender and Equity

- Resource Support
- Technology development

Approach

The organisation builds on the support of and in association with the community government agencies, non-government resource organisation and committed individuals. Awareness, planning and implementation of participatory self-initiatives while working hand-in-hand with the rural and urban communities. The organisation emphasises on participatory initiatives with support of resource, finances, knowledge, information, skills and experience. The organisation has a committed team of socio-econo-technocrats with a varied and right experience of social awareness, health, agriculture, gender, income general activities, etc. The team fabric is reinforced through a strong development perspective.

Area of Operation

The area of operation is Aurangabad and Jalna districts of Maharashtra.

Activities

The urban and rural interventions of Nirman are:

Health

The organisation runs a charitable dispensary since 1994, in a slum area of Aurangabad, The health services of organisations, practitioners and visiting specialist consultant are provided to slum dwellers of Aurangabad city. The rural community also avails these charitable services. Nirman has strengthened its health activities by networking various hospitals, clinics and experts for referral services. At present more than 50 hospital and experts including Government Hospital & Mahatma Gandhi Mission's Hospital, are supporting the social cause by providing all health services free or at concessional charges and free medicines. The rural health extension team supports health awareness and prevention. As of today health needs of 10,000 urban and 900 rural families are addressed. Health check-up and diagnosis camps are organised every three months.

Natural Resource Management

The rural economy is largely based on agriculture, therefore, Nirman initiated its rural development activities with watershed development as a tool of resource management. Nirman in association with Watershed Organisation Trust (WOTR) and National Bank for Agriculture & Rural Development (NABARD) is facilitating participatory watershed development under "Indo-German Watershed Development Programme (IGWDP)."

Nirman believes in the capacities of the villagers and hence, plays a catalytic role in natural resource management. Presently, watershed treatment and resource reclamation is under - taken over 3,360 hectares of government and private land with 571 families in 9 villages. The major benefits as cited by the villagers and observed in impact assessment studies are soil conversation, soil quality enhancement, run-off control, increased recharge, safe and sufficient drinking water, sufficient fodder availability, increased agricultural productivity, better livestock management, enhanced vegetative cover, increased wild-life, strengthening of social structure, etc. Alongwith these increased skills of the community, more income-generation opportunities, better access with the outer world, etc. have developed simultaneously.

Agriculture Support

The livelihood and income pattern are governed by inward financial transactions. As a felt need, Nirman mobilised resources for land capability based better agriculture practices and agriculture credit. Training conducted for land capability based crop planning, improved varieties, dryland horticulture, silvi-culture, agro-based income generation activities, organic and bio-fertilizer management, animal husbandry etc.

Capacity Building and Training

Capacity building of the community and strengthening village institution for participatory resources management and maintenance through learning by doing and structured modules. Promotion of village youth as "Village mobilisers" is also done bt Nirman through proper social and technical, in house and on-job training in participatory planning, implementation, monitoring etc. Exposure visits and in-house training were conducted for the community and village mobilisers to facilitate experience sharing, cross learning etc. These village mobilisers are also carrying further the movement of participatory resource management to many other villages. These advanced skills have helped them to earn their living and status in the society. Alongwith watershed training, Nirman has also organised special trainings for leadership development, health, education, gender development, self help activities and such other needs based trainings.

Community Organisation and Resource Mobilisation

The democratic values and social structure can only be maintained if the community is organised and village institutions are recognised and strengthened. The village institutions such as self-help, groups, panchayat, user's groups, mahila mandals, yuvak mandal, gram sabha, etc. are promoted and strengthened through various efforts. The forward and backward linkages of other institutions such as government departments, bank, PRIs etc. and others like hospitals, subject matter specialists, local entrepreneurs, etc. are also promoted and strengthened, though Nirman.

Nirman's efforts have motivated the villagers to mobilise such resources for School & Panchayat building, road, drinking water facility, loans etc. The villages are capacitated and feel confident of resolving their problems.

Education

The awareness for importance and usefulness of education is realised amongst the villagers over a period of time. Education projects for adult literacy, afternoon classes, non-formal classes for school drop outs are undertaken. Educational motivation activities to create a favourable environment between teachers, students and parents within the present system has enhanced the relationship and attendance at schools.

Support Organisation

Nirman has supported various other organisation with its skills and competence. It has conducted technical and social surveys for soil & water conservation projects of Government of Maharashtra, IGWDP etc. and participated in planning and implementation as technical resource for projects in Maharashtra, Madhya Pradesh, Orissa, Himachal Pradesh, Jammu & Kashmir since last 10 years. The organisation also supports various small organisations in HRD, training on-job experience and institutional support.

Future Plans

The experience and learning of all the social interventions has laid a sound foundation for Nirman. The organisation will further strengthen its field base while acting as a resource organisation. The strategy is to replicate proven technologies & approaches and broaden need-based initiatives. In future Nirman will facilitate.

- Participatory natural resources management projects
- Mainstreaming gender
- Establishing health foundation
- Sustainable agricultural practices
- Local resource based income generation activities
- Resource & Training centre
- Agriculture support
- Micro-finance

Janarth

Janarth is a Sanskrit team which means for the people' was started in 1986, in a chronically drought prone region of Central Maharashtra. The Project aimed at integrated development of the area. Janarth was formed by the coming together of a group of professionals from different fields wanting to take up the challenge of making a perceivable change in the lives of people inhabiting backward regions. The response to this challenge is today translated into action by a team of over 50 staff who mans the different departments of Janarth. To help the Project in its efforts are village level teams of community workers in health, education, economic programmes and relief.

Objectives

- Women empowerment
- Preventive and rehabilitative medical work

- Economic development through income generation programs and marketing facilities
- To emphasise the preventive and promotive aspects of health care
- To promote education
- To enrich the environment by natural resource management
- To empower marginal farmers by maintaining agricultural service centre.

Area of Operation

Located 350 kms northeast of Mumbai, in the Marathwada region, Janarth's area of concentration is at present 44 villages of Gangapur Taluka in Aurangabad District. The weather is dry and the climate extreme, Rainfall averages 18 inches a year but is erratic. Temperatures soar to 43o C in summer and drop to 7o C in winter.

On the strength and experience of their work in Gangapur block they began a satellite project- Janarth - Shahada' in October 1996. The project presently covers around twenty villages of Shahada Taluka in Dhule District of Maharashtra. The villagers belong predominantly to the Bhil and Paora tribes.

The area is poor in basic facilities, thus adding to the reasons for its under-development. When they began their work, only 6 out of 40 villages had all season roads, restricting thereby the state's transport services to these villages alone. In the last few years road construction has picked up, leading to a marginal improvement in communications.

Community Participation

Janarth devides through a continuing dialogue with the villagers, thereby meeting their perceived and articulated needs. The dialogue process also helps in making the people vocal, sensitive, responsive, and willing to take responsibility for their own development.

Activities

Health

Two major foci have emerged in health work -differentnly abled and Ayurveda. Although a part of health sectors, this has grown into an independent department. They began with detection of the differently abled through surveys, health camps and school screening, then worked for their treatment, mainly through government hospitals. Government support has been quite remarkable and helped shape their work in its early stages. Today, many private practitioners too have come forward to help. This has made their work more effective and responsive to immediate needs.

Besides medical rehabilitation and prosthesis distribution they are now able to reach out to the disabled to assist them in fulfilling their social and educational needs. With successful cases to show, it has become comparatively easier to get people to come forward to seek help. Recently they have introduced disability workers training programme to increase the affectivity of their work and to partially fulfil the needs of the region.

JANARTH AGRICULTURE SERVICE OUTLETS

Graph-1

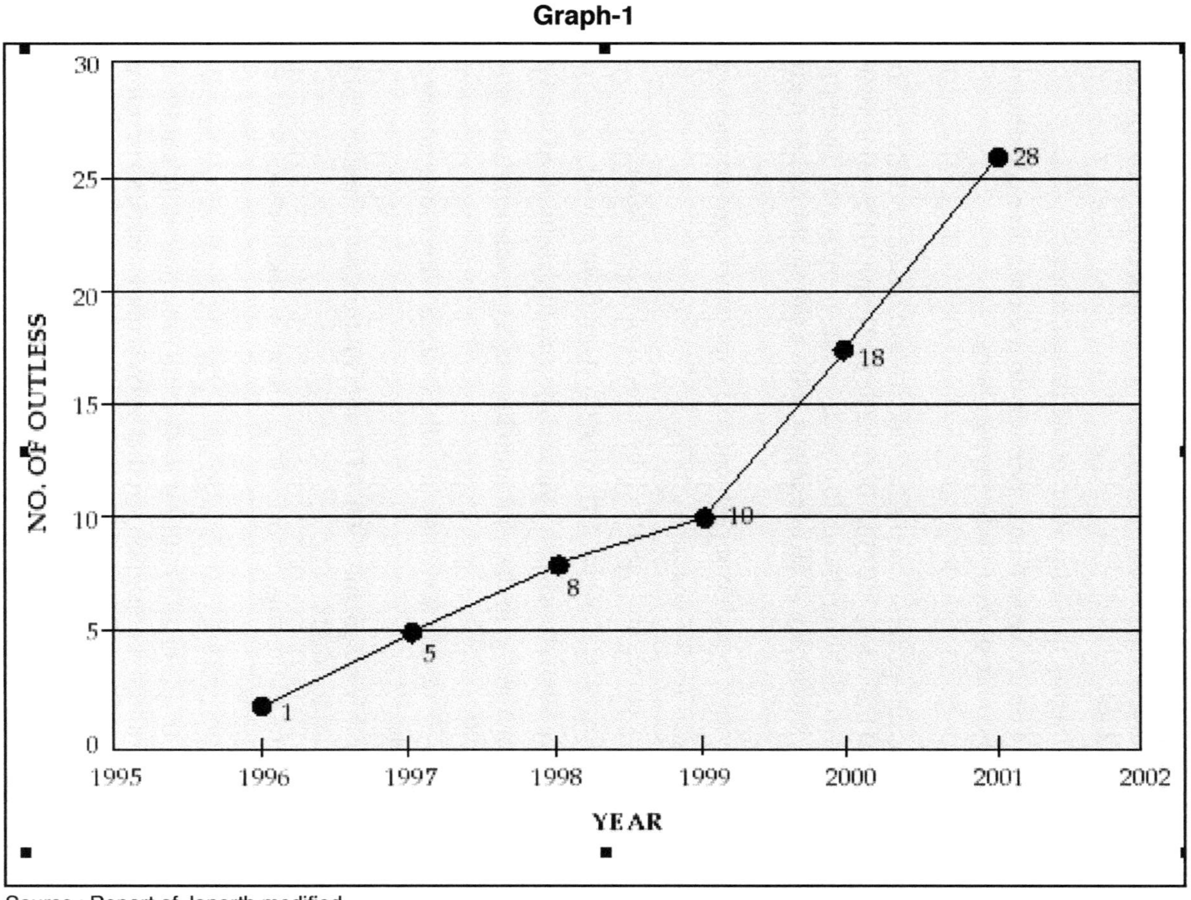

Source : Report of Janarth modified.

Janarth objective here was to re-introduce home remedies in the villages. Towards this end, they found themselves working on a host of programmes -

- Developing an Ayurvedic nursery to ensure that the seeds/saplings they propagate are of confirmed medicinal plant species.
- Conducting meetings of mahila (women)'s groups to increase awareness of the use of Ayurvedic remedies and to impact methods for making these medicines.
- Encouraging the cultivation of medicinal crops.
- Setting up an Ayurvedic Laboratory to test the active constituents of herbs and shelf life of Ayurvedic preparations.

Education

Janarthis multi-programmed educational effort has an underlying and strongly held belief, that literacy is the minimum basic tool required by an individual if he/she is to be given a fighting chance to overcome his/her poverty. Their programmes therefore, attack illiteracy at all age-levels, with a dominant focus on girl-children and women.

Their activities, which complement and supplement government efforts, are:

- Balwadis for 3-6 year age group.
- Study Centres and Sports clubs for school going children.
- Non-formal classes for girls who drop out from school.
- Science laboratories in all the middle and high schools in the Project area.
- Akshar Yatra to make minimum literacy available to adults, through their children.
- Extra-curricular activities in collaboration with Government Secondary schools.
- A monthly magazine "SOBATI", for children, and another, "SHIVAR" for the literate adults.
- Cultural gatherings, sports competitions workshops, exposure visits and excursions.

The Government recognises their efforts and extends help to them whenever requested. Their sports and cultural meets, the only one of its kind in the Taluka have gained enough recognition, so that even schools, which are not covered by us, look forward to participating.

Economic Activities

JanarthisThe economic activities further divided into Agriculture, Income-Generation and Consumption credit support. These activities easily attract the most attention, as they address the immediate survival needs of the villagers. Relevant Government Departments and the Agriculture University in the area, contribute to their efforts in this sector.

Agriculture

Agriculture Service Centres(ASCs) : These are outlets for hiring out agricultural implements and for the sale of agricultural inputs at fair prices. Over 2000 farmers, annually, are members of these Centres.

In 1988 they started with a single outlet with a turnover of around Rs. 1.2 Lakhs. Presently there are in all 26 outlets serving 40 project villages with all except 4 in interior villages, which do not have all season roads as shown in Graph 1. The main beneficiaries are the small marginal farmers. The turnover has crossed Rs. One Crore.

The development objectives served by these outlets are :

- The farmers get assured quality inputs and in time.
- They can purchase as and when they need, at their doorstep.
- They do not need to succumb to unfair trade practices.
- They can avail of agricultural credit of ASC they are members.

Warehousing

This scheme helps farmers who find themselves in a cash crunch, enabling them to get money against the grains they store in the warehouse. This prevents distress sales of their produce.

Warehousing and Marketing Assistance : On the insistence of farmers in 1996 they launched this activity as a logical next step to ASCs and Warehousing. While ASCs aimed at improving the yield, this activity aimed at getting the farmer a better price for this yield. To influence the local market conditions] which include price for their yield in favour of the poor they made a modest beginning in two local markets in the project area. The turnover has crossed Rs. One Crore. Their experience over the last year has shown that they can influence the local price between 2-8% and also help avoid distress sale by farmers. The response has been very encouraging.

Income generation

Small Loans are given to groups or to individuals to enable them to generate additional income with the help of skills at their command. For example : trading, repair shops, tailoring, poultry etc.

Land Development

Under this, group loans are given to farmers for repairs of wells, for irrigation pipes, electric motors, land repairs and contouring, all aimed at increasing the yield from the land.

Natural Resource Management

In this area, their effort is aimed at exercising the physical environment by conservation measures and through the efficient use of natural resources. They also believe that if they irresponsibly only take from nature without giving back to it, they will be working towards their detriment. It includes following activities:

- **Land Contouring** prevents soil erosion by guiding the flow of surface water run-off along the natural slopes of the land, and by checking its speed.
- **Gully plugging** repairs the land while preventing further soil on caused due to fast flowing rain water.
- **Check-dams** help in collecting surface rainwater which then gradually percolates and re-charge the ground water. They also study the impact of these dams by measuring, the water levels in wells around the dams monthly.
- **Plant Nurseries** subsequent tree plantation help in giving back tree over to the land. This in turn helps by giving fuel, fodder and fruits. This activity is carried out with the help of the Social Forestry Department of the Government.

Government DPAP programme

In 1996 the agency was selected under the Government DPAP programme for Watershed Development. Eight watersheds have been sanctioned. This includes land treatment and water conservation works. The emphasis was involving people in every stage of work through registered village committees.

Relief Work

This necessarily is an ad-hoc and need-based intervention. In 1995 Janarthtook up a plague prevention programme on a war footing. They responded to the Latur quake in 1994 and carried out yearlong rehabilitation work in 2 villages. In 1992 when many houses collapsed due to heavy rains in their area they responded with providing tin sheets for roofing. And, in two drought years, they have provided cattle-feed, wages for work and grains to the very poor.

Two activities, which began as relief interventions, are now run as regular programmes because of their need and relevance - Khawati, and Contouring and Land Repair work.

Khawati

This is a grain loan given in the lean months, to needy landless labourers and marginal farmers. Over 1000 families from 30 villages benefit each year. This programme keeps the families safe from the usurious rate of interest charged by the moneylenders. For some, it also protects them against the need to migrate. In the Khawati programme, which necessarily attracts the poorest, their rate on the loans disbursed is over 80%.

Women empowerment

The agency encourages the involvement of women in all their programmes - both at the planning and the implementation stages. They pursued this in many unobtrusive ways - a policy which expects both spouses to sign on an agriculture loan agreement; small income generation loans given predominantly to women's groups; economic programmes in some villages given over to the women to run;

women representatives on committees and meetings etc. Now that they have gained the confidence of the people, they have programmes that are more visibly and exclusively targeted towards women.

An awareness building activity that has formalized into a full fledged programme is the Mahila Wing activity. The aim of their women's wing is to build confidence in women through increasing their information/knowledge base, and thereby their participation in all dimensions of village life that affects them. Programmes taken up under this are :

- Forming of Mahila mandals and organising cultural programmes for women.
- Incentive schemes for participation in mandal activities include, seed gathering, medicinal nursery raising and collection of medicinal raw material.
- Conducting workshops for women and girls on their legal rights.
- Aiding effective participation of women in village governance through their elected representatives. The various activities under this are - holding workshops, giving the elected members their status in village meetings, printing training material, organising exposure visits, etc. Women are getting together in informal groups to discuss common problems.

Urban Slum Activities

This activity was introduced by Janarth in response to the need felt by their well wishe, that Janarth have an urban presence. In their own thinking they justified their presence by initially working among the poorest of the poor - the rag pickers. They have started their intervention in 3 slums on the periphery of Aurangabad. Their activities include:

- Ayurvedic clinic & health camps.
- Balwadis and Sports groups.
- Liasing with the Municipal Corporation for the provision of adequate civic amenities.

Abhinav Vikas Sanstha (AVS)

The founders of Abhinav Vikas Sanstha] AVS learnt that promotion of self-sustained and participatory intervention lays the foundation for socio-economic development. AVS imbibed the experimental learning of its members within the organization and it aims at integrated development through need based intervention[10].

Objectives

- Awareness and mobilization
- People's initiatives
- Natural resource management

- Gender sensitization
- Skill and capacity building
- Credit linkages
- Agriculture promotion

Approach

"Development is a continuous process facilitated through committed and cautions efforts" evolved as the objective and approach of AVS. AVS plans to achieve through various developmental programs and activities with the following approach.

- Awareness building and community organization
- Need assessment and prioritization
- Capacity building and training
- Participatory planning and implementation
- Social Empowerment
- Capability based resource use
- Equity and sharing

Area of Operation

Presently, the organization focuses its efforts in Soegaon tehsil, the most backward block of Aurangabad district. Soegaon lies along the northern boundary of the district marked with hilly undulating terrain along the Satmala and Ajanta hills. The area experiences extreme weather conditions. Due to uneven rainfall the area faces scarcity and drought, very often.

The total area of tehsil is 689.43 s.q.k.m., distributed in 79 villages with total population of 77,213 covering 15,243 household. The tehsil has highest population of Scheduled Castes and Scheduled Tribes in district, 10.39% and 11.69% respectively, of the total population. The literacy rate of tehsil is 48.03% and is completely rural and the only tehsil is 48.03% and is completely rural and the only tehsil without urban population. Despite being block headquarter, Soegaon is locally governed by a Gram Panchayat.

The land distribution is uneven and concentrated in the hands of big landlords. 49.56% of the total land holders are small and marginal farmers holding only 20% lands and 2.3% are big land lords who possess 10%. Area owned by forest department is 20% of the total area and falls in about 60% villages.

There is only 1 government hospital, a dispensary and 3 PHC's in the tehsil. Annual growth rate of this tehsil is 2.4%, lowest in the district, mostly due to lack of medical facilities. Only 50% villages are connected with roads.

The area is marked with poverty and backwardness, unjust societal structure, low literacy among women, depleted resources regime. Major problems and needs of the area are -

- Degradation of natural resources
- Unemployment
- Lack of irrigation facilities
- Lack of awareness and empowerment
- Exploitative status of women
- Lack of information and knowledge

Activities

In view of objectives, aims and approaches laid down and the scope for development, AVS has facilitated community-based interventions in 6 villages of two clusters from 1998. Self-help initiative lead to self-decision and empowerment, therefore, AVS promoted common interest groups (CIG) in cluster villages. These activities are voluntary. Natural resource management, participatory irrigation management, etc. are supported through various agencies.

SHG and Women's Development

Initially, group formation and community organization was initiated with the most deprived class, women. Youth and farmers were further promoted to form self-help groups. At present in 6 villages there are 30 SHGs of 387 members and their total savings is Rs. 4.0 lakhs. The groups meet regularly, share their feelings and have established their own Identity.

Krushi Mandal

Krushi Mandal serves as a platform for information sharing and transfer of knowledge for agriculture promotion. The farmers interact with experts from various S&T institutions and are benefited by modern technologies.

Youth's Activities

Youths are future face and centre of entire community. AVS concentrated hard to organize this human resources, to involve them in mainstream. Youth groups are promoted to address their problems. AVS will promote these groups and liaison with district administration. The groups participate voluntarily and have constructed a bus stop costing Rs. 15,000 at Soegaon through shramdan. With the help of Youth SHG 5 members have establish their own business.

Natural Resource Management

Watershed development project for Varhkedi was undertaken under Indo-German Watershed Development Programme in 1999. Due to degraded resources, drought conditions and lack of agriculture facilities the community resorted to temporary migration as sugarcane cutting or construction labour, to Mumbai and Surat. Scientific and integrated watershed development approach with emphasis on soild and water conversion and biomass regeneration was adopted. Components of watershed development project are;

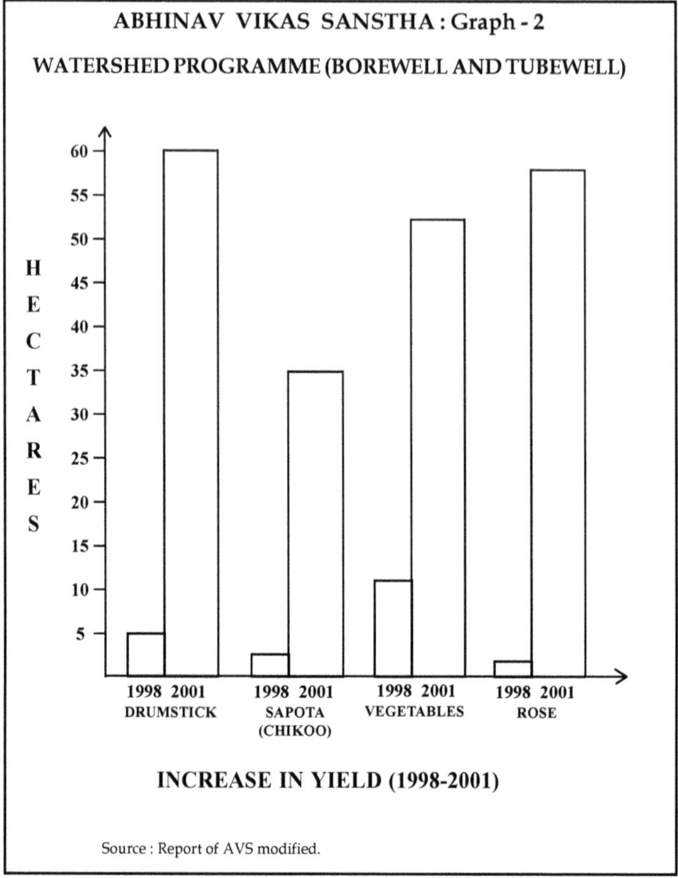

Source : Report of AVS modified.

- Community Empowerment
- Environmental Systems
- Soil and Water Conservation
- Production Systems

Optimal Use Demonstration

Watershed development provides necessary capacity and infrastructure for land based activities. Since, the community was unaware of modern technologies and improved agricultural practices and were not confident of its benefits, AVS promoted on-field demonstration of these technologies, on pilot scale, for optimal use of land and water resources. Demonstration aimed at -

- Land management
- Change in cropping pattern
- Water management

On-field Demonstrations by AVS for;

- Horticulture - drumstick and sapota
- Floriculture - rose farming
- Traditional crop
- Vegetable crop
- Pasture land and tube well
- Micro-irrigation

Drumstick and Sapota are low water requirement plants and less labour intensive, Rose farming is more economical than cotton. Alongwith these some area was taken for comparing economics of traditional crop over horticulture and floriculture. The demonstration has boosted the confidence of farmers and 29 new tubewells were drilled leading to irrigation for 60 hectare with a net gain of 100%, a change in cropping pattern and adoption of modern technology (Graph-2). Therefore this demonstration has served as a milestone for the than poor farmers for their economic development.

Micro Finance

Modern Day agriculture requires external inputs, which are mostly not affordable to poor farmers of the area, therefore a revolving bridge fund is made available to marginal and poor farmers as soft loan, through AVS. Farmers can avail this facility without legal formalities with an interest rate of 2% per month, and will be revolved within the community. These funds can be availed for following activities ;

- Dairy promotion
- Horticulture
- Forward linkages
- Agricultural support
- Skill training
- Information Centre

Grasp

Grasp was created to professionalise the process of development at the grassroots by providing an interactive platform for the stakeholders. The vision of grasp is empowerment of communities to manage their resources in a professional and environmentally sound manner and to achieve this, their mission is to work for socio-economic development of the poor through community action, capacity building of grassroots level organisations, institution building, policy advocacy and action oriented research[11].

Objectives

- To develop strategic focus in identifying the technical and organisational management needs of grassroots level organisations.

- To strengthen the management and utilisation of the natural resources and to regenerate the degraded lands to fulfil the demands of the local community.
- To build capacities of the grassroots level organisations and CBOs and promote new grassroots organisations to work for ERM.
- To develop areas and regions plan to implement and replicate the WDP/ ERM programmes.
- To develop the strategy for research, network building and advocating the policy for broad basing the concept of community empowerment using ERM as a tool.

Area of operation

Currently grasp focuses on Marathwada region (7 districts) and Vidharba region of Maharashtra State and Southern part of Himachal Pradesh.

Activities

Grasp with its professionalism mode promote sustainable system while combining its professional competence with social commitment.

It's Community Action Projects (CAP) alleviates poverty while protecting the environment in an integrated watershed framework with an emphasis on holistic development of the region. Working with over 3,000 families and regenerating 12,000 Ha of degraded land (which includes 3,000 Ha of forestland). (Table-4).

- Village watershed commitees (VWCs) being formed focuses on building local capacities to maintain the resources created and work on forward linkages.
- Mahila mandals (MMs) initiated to increase direct participation of women in economic and developmental programme.
- Community banking system-to provide micro-credit to rural poor and a consecutive step towards empowerment of women as a cohesive force.
- Forest Protection Committees (FPC) have been formed to save the resources by controlling biotic interference under joint Forest Management (JEM) act.
- It's Development Initiative Support Programme (DISP) anchors the philosophy of building locally based organisations and creating new organisations to replicate the work in hinterland. These NGO's focuses on natural resource management, women empowerment, micro-finance, preventive and curative aspects of health and education. Presently working with over 5000 families for holistic development of the area.
- It's Team for Research Evaluation and Development (TREAD) develops research strategy, foster institutional linkages through networking and advocate policy level decisions. It undertakes potential related studies and prepares regional strategic plans for optimum utilisation of natural resources.

- Guild of Regional Associates for Community Empowerment (GRACE) is a network of NGOs initiated by grasp with a prime objective to provide a common platform for sharing of experiences, capacity building and to broad base the concept of community empowerment.
- Grasp has worked for increasing focus on the community participation in the management of the environmental resources through planned activities, execution of projects by the community groups and adopting the Watershed Development Approach as the key tool to achieve the stated objectives. The professionals were committed and has varied and vast experience in understanding and analysing local need and institutionalising it for making the community and henceforth region self-sustainable.
- GRASP developed approaches to foster the land and water management initiatives for the rural communities specially the poor. grasp team approached the community by initiating a need-based dialogue with the village community on specific themes/ issues, such as self-help groups formation, installing micro-irrigation schemes in groups, soil conservation works, water recharging structures, provision of improved agricultural packages, enhancing livelihoods through optimum utilisation of land and water and other related issues. Team focused on provision of technical and organisational support to the community/groups. Project is implemented in hands on manner to inspire, train and support local initiatives by local NGOs. Community based groups to maximise optimum utilisation of land and water. grasp succeeded in building, both formal and informal, relationship with Community Based Organisations.
- Grasp performed a number of DPAP programmes with the government (Table-5).

Anti-Alcoholism Movement

Grasp educated the women folk the hazards of alcohol. The women organised under the Women group] Mahila Mandal leadership and took up anti-alcoholism drive. Alcohol brewing has been stopped and alcohol drinking has been controlled to a large extent. This was possible only because the women organised themselves for the fight against social evil.

Social Fencing

The major reasons for the after effects of low rains in drought prone areas are deforestation, soil erosion and excess run-off. The complementing measure for technical treatments for sustainability is social fencing for uncontrolled tree falling and uncontrolled grazing. The village watershed committees and FPCs have taken up the responsibilities.

Joint Forest Management

Grasp under the IGWDP involved villagers in forest treatment and management. Forest Protection Committees have been formed with consensus and forest area of

315 hectares has been treated. Though the government agrees principally and has drawn up the plan for the same, actual implementations have hundred hurdles. The Joint Forest Management (JFM) practice has taken shape only through continuous efforts and follow-up by villagers, FPCs and Grasp.

Grasp Watershed Programme
With Community Action Projects (Table-4)

S.No.	Watershed	No. of Villages	No. of Hectares	No. of Families	People's Contribution
1	Bottek	6	1345	179	
2	Sitanaik Tanda	2	1685	165	
3	Hatmali	2	1195	684	Totallng to
4	Naigavan	1	1766	703	more than
5	Donwada	1	774	312	Six lakh INR
6	Borwadi	1	652	248	
7	Waghola	2	1241	286	
8	Babra	1	3075	1047	
9	Kankora	1	5	2	

Source : Report of GRASP modified.

Grasp Direct Implementation Programme
In Collaboration With Government (Table-5)

S.No.	Watershed	Programme	Area	Families
1	Hatmali	DPAP	1995	684
2	Naigavan	DPAP	1766	703
3	Borwadi	DPAP	652	248
4	Donwada	DPAP	774	312
5	Waghola	Adarsh Gaon	1241	286

Source : Report of GRASP modified.

Migration

The whole of the village migrated to distant places in Maharashtra and Gujarat around 500 km away. These migrations had severe effects on health, education and social structure of the community. Since the implementation, group programms, migration decreased drastically from 95% to 5% and some of the families who had migrated permanently have started returning to the village.

Wild Life

Soil conservation treatments, grass development and social fencing have today enhanced vegetative cover and changed the scenario of the forest area, which has attracted the wildlife.

Integrated Health Management Pachod (IHMP)

IHMP is a trust registered under the name Ashish Gram Rachana Trust on 17th March 1979. Its training centre and main office is located in Pachod, Aurangabad district, Maharashtra. It established an urban office in Pune in 1995. Now, IHMP has shifted to Pune office.

Area of Operation

IHMP has been working in the underdeveloped Marathwada region of Maharashtra for the past 20 years. Today the agency is working in 72 villages of Paithan Taluka with 100,000 population and 27 slums of Pune city with a population of 30,000.

Objectives

IHMP has adopted a life cycle approach for program implementation. Information, education and communication (IEC) are core programs of IHMP, it aims at;

- The holistic development of the individual, family and community.
- Upliftment of marginalised groups.
- The health and development of women, children and most disadvantaged.

Mobolising communities toward self-reliance and sustainability and to organise children and adolescents, through development of health, water and sanitation programs.

Organisation

All the programs and activities of IHMP are implemented through the executive body of the institute. The institute has two conference halls, hostel for up-to 32 trainees, residential facilities for external faculty, laboratory, library, documentation centre with photocopying facilities, audio-video library.

Activities

Primary Health Care

It aims at generating a demand for quality health services for women and children. It operanationalises primary health services through participatory community assessment. It provides those services that are not available through the government health system such as womens health, gynaecological morbidity, STD/HIV prevention, neonatal care etc.

Bal Vikas : Child Centred Development

This promotes development of children through chlid-centred approach. It places the children in the role of change agents working towards the development of their village. Through this process, children develop themselves also. Bal Panchayats have been organised for mobilising and involving children in various developmental activities. Children provide non-formal education to non-school going children in their villages. They also undertake environmental projects.

Navjeevan Rugnalaya

It is a 30 bed hospital which serves as a referral centre for villages within a radius of 30 kms. Its primary focus is on women and children.

Life Skills for Adolescents

This program addresses the special needs of adolescent girls and equips them with life skills for a better quality of life. The program also focuses on sexuality education, anaemia prevention and improvement of the reproductive health of adolescent girls. The girls have gained self-confidence, legal literacy, nutrition and health.

Environmental Issues

IHMP being situated in one of the most drought-prone regions has necessitated its involvement in environmental issues such as water, sanitation and afforestation. The institute has decades experience in rural drinking water supply and has developed systems for community based management of hand pumps. The programs include repair and maintenance of hand pumps; community based monitoring; water quality testing, health education and community organization.

Disaster Management

The institute has extensive experience in earth quake related disaster management. This includes disaster related relief, rehabilitation and reconstruction work. The institute works with a level of community participation in its rehabilitation efforts. The institute was also involved with the relief work in Latur district and Gujrat in 2001.

Policy Advocacy

IHMP's policy initiatives have been at local, state and national levels. The areas where IHMP has had a big impact on government as well as NGO policy are child health and nutrition, adolescent health, reproductive and child health, water and sanitation etc. IHMP works with various networks for policy change and advocacy. IHMP has a centre at Pune developed for low cost audio-visual and training materials for the NGO and government sectors; and for addressing urban health issues in slums. This centre also has a research and documentation unit.

Dilasa

Dilasa Janvikas Pratishthan is a voluntary organization working in Aurangabad District of Marathwada. It is engaged in the socio-economic uplift of villages and ameliorating the standard of the women in family and the society. It works with two target zones in the mind viz. watershed and women. Dilasa plays a catalytic role in socio-economic development of villages by co-ordination of various government agencies and the villages together. It helps the people of the village to become self-reliant by assisting them in income generating activities on government various development schemes[13].

Objectives

- The prime objectives of Dilasa Janvikas Prtishthan is women empowerment and self-reliant, self sustainable development of villages. The organisation works for these objectives through;
- Development of women and children in rural areas (DWACRA)- focussing on the women living below poverty line
- Ecological concerns and conservation and management of natural resources and forestry
- Land capability based agricultural development
- School sanitation
- Resource support and technology development
- Gender equity and education of women

Area of operation

Dilasa works in the villages of Aurangabad district of Marathwada.

Activities

Implementation and Monitoring of DWCRA

Dilasa is engaged in this subscheme of IRDP in Aurangabad District since last three years with the primary objective of focusing attention on women of rural families below poverty line with a view to provide them opportunities of sustainable self-employment. Dilasa formed 310 groups of 10 to 15 women each for taking up economic activities suited to their skill, aptitude and the local conditions.

Rain Water-Harvesting Structures

Dilasa works with Unicef in Rainwater Harvesting Schemes (RHWS). It has developed and adopted a unique modern technical experiment to harvest rainwater in specially made tanks through inclined channels from top of the roofs through filter and other assembly. The highlight of this program was that these 104 structures built in Aurangabad district were built by women masons, proving women can do any type of job if there is proper training and encouragement.

Watershed Development

Dilasa strongly believes that watershed development is a key of the rural development.

Dilasa undertakes watershed development projects supported by Indo-Germen, Unicef and NABARD. The project areas were Kachheghati (Tal.A'bad), Murumkheda (Jalna) and Karhol (Tal. A'bad) are the Indo-German projects. The project areas of Unicef are Nagunichiwadi and Darakwadi (A'bad) and the project areas under NABARD are Jalkotwadi (Tal. Tuljapur), Manmodi (Tal.Tuljapur) and Aliyabad (Tal. Tuljapur). Dilasa is also implementing watershed programs under DPAP.

Village Eco-Development Project of Social Forestry

This is called VED project, started in 1997 to take up the challenge in totally neglected village. It is located at the corner of Aurangabad Taluka. The results of the program are very inspiring. People have taken initiative to grow variety of grasses on the top of hillock, due to which plenty of grass seed is available with them which is further requirement for the grass seeding on the bunds.

Smokeless Chulla

Dilasa has undertaken this project for the emancipation of women. This project has been completed in following villages; Hirapuri/ Shamwadi (132 chullas), Peerwadi (71 chullas), Kachheghati/ Murumkheda (250 chullas) and Konewadi (77 chullas).

Vermiculture

Dilasa undertaken this new challenging project with Unicef. It has constructed vermipites in Karanjgaon, Konewadi and Murumkheda, Dhavla puri,etc with active participation of the farmers. Dilasa is looking forward to implement vermiculture project for the whole of Aurangabad Tehsil.

School Sanitation

Dilasa is implementing the sanitation program in 218 schools in Aurangabad Taluka. It believes that this program of sanitation covers aspects of environmental and household cleanliness as well as hygiene.

Multipurpose Women's Centre

Dilasa has established this centre with the objective to give legal aid to women, family counselling, and income generating activities and rural entrepreneurship.

There are trained employees to handle various issues of women. Since 1995 this centre handled 200 cases of women, provided training for SHG's, dairy, nursery, health camps.

Marathwada Sheti Sahayya Mandal (MSSM)

Marathwada Sheti Sahayya Mandal (MSSM) is a pioneering Voluntary Organisation working in Central Maharashtra for rural development, in general, and watershed development in particular. Started as a public charitable trust to assist small and marginal farmers to improve their living conditions through increased agricultural production and productivity, MSSM undertook several developmental activities in land and water management sector in the last three decades. MSSM is a voluntary agency registered as a trust and a society in Aurangabad under the Bombay Public Trust Act February. 10, 1969 and societies Registration Act. Feb 4, 1969[14].

The church of Scotland had been working in and around the village of Jalna, trying to alleviates the problems of the lack of agricultural development causing mal development, drought like situations in this area. The huge developmental need could not to be shouldered by a single organisation. So the joint effect of

Scottish Committee for War on Want, Christian Aid of Britain and Oxfam of Britain and Church of Scotland started visiting these areas for development of surface and underground water resources, which is the key to all development.

The MSSM was formed in 1960, in collaboration of War on Want department, church of Scotland Mission, the mandal, which constituted of a governing a working committee elected under the rules and regulations of the mandal, worked together for the integration and development, and expansion of development of rural Maharashtra. The philanthropic, humanitarian, non-political, institutionalized secular outlook of MSSM as a voluntary organisation was formed.

Objectives

- To provide, render, endow, help and assist in all and every manner, for the uplift and welfare of farmers, farm-workers, poultry-farmers, dairy-farmers including all those connected with agriculture & its allied industries; needing economic and other support to attain better living and to increase their standard of living by increasing productivity and efficiency in their profession.
- To conduct research in subjects of agriculture and its allied industries by creating and managing such centres of establishments or by aiding such institutions.
- To impart training to farmers, farm-workers, poultry farm-workers, dairy-farmers including all those connected with agriculture by creating and managing training centres, institutions, conducting course of training, maintaining libraries in the subjects of agriculture and its allied industries and to impart knowledge and development to the above mentioned sections of the society and to aid institutions conducting the above activities.
- To establish, create, conduct, maintain or aid service organisations among others for rendering services of various nature with the object that the farmers, farm-workers, poultry-farm-workers, dairy-farmers including all those connected with agriculture are enabled to take advantage of such machinery and services, which are beyond their reach.
- To establish, create, manage or aid Panjra-pol or Gaushala and or breeding stations of improved cattle and poultry[15].

Vision of the Agency

MSSM started as a public trust to assist small and marginal farmers of the region to improve their living conditions through increased agricultural production and productivity. MSSM was, although working mainly for agricultural development, involved all along in searching technological options and devices for managing farm-level problems of depleting groundwater and degrading environment. The vision of MSSM, as it evolved over the years, has been to enable the community to improve the quality of life through systematic management of their own resources.

The mission statement of the organisation is "Overall improvement in physical, social, economic, structural and behavioral quality of life through vulnerability reduction and enabling men and women to plan, develop and manage their resources".

Area of Operation

MSSM is presently working in five districts, namely, Aurangabad, Jalna, Beed and Osmanabad of Marathwada, and Buldhana of Vidarbha regions. MSSM has a strong operational base in districts of Aurangabad and Jalna, where they have been implementing several projects in livelihood and NRM sector for over three decades. It is in these districts the foundations of watershed development concept and projects were laid. For development of appropriate agricultural technology and its extension, MSSM is effectively running a Krishi Vigyan Kendra for district Jalna from 1994. From last year, MSSM has been working towards capacity building of NGOs implementing State sponsored watershed development programmes, as a Mother NGO in four districts.

Short Description of major projects undertaken and the achievements MSSM has successfully implemented in the recent past are as follows :

Watershed Development Projects

- Jadgaon Micro-watershed Project (1992-98) in Taluka and District Aurangabad, supported by EZE-CA under AFPRO-EZE-CA Package-IV Programme. The project covered a population of 1811 (295 families) and an area of 797 ha. Total Cost: Rs.68,04,088 including local contribution of Rs.27,50,637.
- Jadgaon Group Watershed Project (1993-99) in Taluka and District Aurangabad, supported by NABARD-KfW under Indo-German Watershed Development Programme. The project covered a population of 2,541 (414 families) and an area of 2,200 ha. Total Cost Rs.1,42,53,538 including local contribution of Rs.11,25,728.
- Asarkheda Watershed Project (1997-2001) in Taluka Badnapur District Jalna, supported by NABARD-KfW under Indo-German Watershed Development Programme. The project covered a population of 1,245 (212 families) and an area of 875 ha. Total Cost Rs.52,36,414, including local contribution of Rs.5,04,172.
- Kadwachi Watershed Project (1998 -2000) in Taluka Badnapur District Jalna, supported NABARD-KfW under Indo-German Watershed Development Programme. The project covered a population of 1,954 (411 families) and an area of 1,541 ha. Total Cost Rs.1,35,68,697, including local contribution of Rs.5,04,172.
- Satana Watershed Project (1998 -2000) in Taluka and District Aurangabd. Initial work of community organisation and planning completed. The project covered a population of 1,434 (222 families) and an area of 1,080 ha. Total estimated cost: Rs.1,13,25,483 including local contribution of Rs.6,79,488. The project is under consideration of Adarsha Gaon Society of Government of Maharashtra.

- Watershed Development in Karmad and other 5 villages (1998 -2000) in Taluka and District Aurangabad. Initial work of community organisation and planning completed. Possibility of mobilising support from Government and other sources is being explored.

Appropriate Technology and Extension

Integrated Pest Management (1999) has been initiated in village Butkheda of District Jalna on 118.5 ha of cotton crop belonging to 64 farmers. The project is supported by Swiss Agency for Development Cooperation under AFPRO-SDC Phase IV NRM Programme. Total cost is Rs.2,332,126, including Rs.1,233,250 as farmers' contribution. After the success in the first year, the practices are now extended to about 200 ha of cotton crop in three other villages, mostly at farmers' own expenses.

Natural Pest Management Programme (1998 - 2001) has so far covered 70 ha belonging to 89 farmers in 2 villages of District Jalna. Supported by Centre for World Solidarity, the total cost is Rs.180,250. In addition, there is substantial contribution of local farmers.

Awareness and training programmes on a variety of topics are being conducted in villages and on campus by Krishi Vigyan Kendra, Jalna.

Training of Rural Youth

- Training in Vocational Skills (1998) has been offered to about 60 rural youth every year with the support from Functional and Vocational Training Forum The project cost is Rs.12,61,000, including local contribution of Rs. 2,61,000.
- Training of Adolescent Girls (1999) has been initiated in the watershed projects aiming at personality development and imparting life skills among adolescent girls. Mostly run with local support and own funds, some financial support has recently been secured from NABARD under Women's Development Component of IGWDP.

Krishi Vigyan Kendra - Jalna (KVKJ)

In the field of training and extension, MSSM had been working towards awareness-generation and motivation of farmers in the regions on appropriate technology for agriculture and sustainable resource management. These efforts, in the initial stages, were sporadic and limited in scale. From 1994, MSSM went into training, education and extension in an organised and systematic manner with the medium of Krishi Vigyan Kendra at Jalna. Krishi Vigyan Kendra Jalna (KVKJ) was established by MSSM for promotion and extension of appropriate agricultural technology in District Jalna, training and demonstration being the key elements in the extension strategy. KVKJ has focussed on promotion and extension of, inter alia, watershed development technology, women's development, farm-processing, non-farm activities for income generation, to attempt sustainability of land and water systems in selected villages. KVKJ has done a pioneering work in extension of Integrated Pest Management technology, particularly in cotton and red gram, in the district.

KVKJ has also been running regular courses, for last three years, in the discipline of agriculture under Yashwantrao Chavhan Open University of Maharashtra. These courses have benefited the youth, mainly school drop-outs or those who could not continue their schooling due to economic and other compulsions, from the rural areas in the district. Four rural youth successfully completing their Bachelor of Science degree course speaks of the achievement of KVKJ in spreading technical education in rural areas.

KVKJ has been running the certificate courses on watershed development and on dairy technology for last one year. These courses are affiliated to Maharashtra Council of Agricultural Education and Research. MSSM has also started organising technical and vocational courses on industrial disciplines for rural youth in the region. Vocational training on appropriate technical trades for school drop-outs from projects villages are being contemplated at present. Training of representatives and functionaries of Panchayati Raj institutions is also on the agenda of the organisation[17].

MSSM looked closely at the gender aspects after the evaluation of Adgaon project, which revealed inequitous of benefits among men and women. The men gained a great deal more than women on technological advances and the women did not advance even marginally on technological front. Such a finding made MSSM to re-look at its interventions and strategy in order to address the issue of differential benefits.

At the organisation level, it was realised that the earlier approach of addressing the head of the household , a man, and assuming that he represents the concerns and interest of all' family members, including women, was not correct. Therefore, MSSM initiated measures to address women separately and independently. In their participation in development activities. Although these were beliefs and opinions, which are continuously reinforced through repeated patterns:

- Women's role was seen mostly in the domestic sphere, and they were not believed to possess any ability to be in the social or public domain.
- Women were traditionally not given an opportunity to show their talents in public or social activities, and thus were not encouraged to take part in development activities.
- Women were not believed to have any knowledge of land and water or watershed, and hence, were not considered to be in a position to contribute to planning.

This followed several small activities addressed to reduce the drudgery of women, viz., kitchen garden, mandals and savings and credit groups to help them out their sundry credit requirements, as also to help them gather skills and aptitude to manage economic activities. These were useful in overcoming the initial inhibitions in working togeher of both the organisation and the women. It also helped to see the limitations of working with women alone in isolation of the mainstream development work.

MSSM attempted to build upon the lessons and to focus the efforts on gender - and not narrowly on women alone. In subsequent watershed projects, MSSM made some simple interventions towards both women's empowerment and integrating gender. Some of these were systematically planned while some others were not. The planning was usually done in a series of meetings with mahila mandals and watershed committees. Normally the ideas emerged during the discussions on ongoing activities and on their expansion or follow-up. Ideas developed into workable plan were discussed with the watershed committee for finalisation. The process usually took 3-4 months. Other organisations working closely with MSSM and grasp followed the footsteps, and the interaction helped in refining the methods gradually. Developing the capacity of the human resources and field workers was also attempted in the process.

Major Interventions for women Empowerment & Gender Equity

The interventions were towards awareness and skill-building and empowerment of women as concurrent strategy and a pre-requisite. Simultaneously, efforts were made in institutional field to bring together men and women and involve them meaningfully in various activities.

Awareness and Skill-Development

1. Specific training programmes, e.g. fertiliser application, pest control, weed control, etc.
2. Involvement in broad or overall village development planning through PRA.
3. Net planning - planning development of one's farm using simple techniques.
4. Gully plug construction.
5. Concrete mixer operator.
6. Women supervisors (in neighbouring district)
7. Bicycle riding.

Socio-Economic Developments

1. SHGs and linking with the banks
2. Tree plantations incentive scheme.
3. Kitchen garden
4. Dal mill
5. Garlic trading

Institutional Field - Towards Integration

1. Setting up of farmer's clubs
2. Training on PRI functioning
3. Watershed Committees and gram sabha.

While some of these were one-off events, the others unfolded into meaningful and useful activities. Some could sustain for a long time, while in case of the others the organisation could not sustain the interest of either the women or the people in general. Some of these are explained in detail in the following pages.

Training on Technology

Initially, efforts were made to train the women on "unusual" skills or techniques not falling in traditional roles of women like concrete mixer operator and gully plug construction. However, these did not yield any significant results except that they were appreciated by the villagers, and like flash in the pan, forgotten. Attempts were made to train the women in supervision skills, but they never practiced these skills. Thereafter, the training inputs were about the traditional roles of women like weeding and fertilizer application.

Fertilizer application training was preceded by a village level seminar, wherein the women also participated and the farmers interacted with scientists on agronomic practices of certain crops. As a result women started discussing, within family and with others, on timing and costs of various operations, as also influencing the decisions.

Pest control training was part of larger integrated pest management programme on cotton in Jalna district. Although the role of scouting of insects (measurement) was with youth, the women and children were educated on the identification of insects, friends and foes, and their control. The result was that the women became familiar with the control measures and on the insecticides to be used at various stages.

Special skills on watershed planning were imparted through a series of practical training sessions on net planning - a method involved by MSSM to plan soil and water conservation treatment by involving farmer. In one village, the women and men were involved in the planning exercise, with result that the women's participation in VWC meetings and Gram Sabhas increased considerably. Further, it demonstrated that some women have better knowledge of their farms and have a good understanding of how their farmland could be improved. This method is now being tried by grasp and Nirman in their watershed projects.

Socio-Economic Development

Although efforts were made to promote income-generating activities for women, the response was limited, because the identified activities fell largely outside the areas and interest familiar to them. Efforts were made to train women on nursery raising, budding and grafting, but the latter's utility remained transitory as the demand gradually reduced. So, MSSM decided to go in small steps and promoted savings and credit groups. Most groups were eventually linked to the banks, which provided the opportunity for women to interact with institutions and also to understand the economics business statistics. They proved to be strong primary units for building mahila mandals or village committees and for launching income-generating activities. A dal mill and collective trading of garlic are being planned in Asarkheda and Kadwanchi watershed.

Likewise, kitchen garden and tree plantation was promoted through women. A scheme was designed to provide incentive to farmer for protecting and tendering the saplings on their own farms. Initially, the results were poor. But later on when the scheme was implemented through the women, there was a marked improvement. One main reason for the success was the scale the women knew exactly how many saplings they would be able to take care of and accordingly decided to rear only that many. This experience was useful in the gram sabha and committee meetings for making many other operational decisions.

Institutional Fields

This was the area where conscious efforts were made to bring together the men and women in a systematic manner. Several small steps were taken to link men and women to the mainstream development activities. The efforts started with eliciting support from men folk and the watershed committee for women's activities. The representation of women in the watershed committee was made through SCGs and mahila mandals. The committees were asked to sponsor partly or largely many events of women. Similarly, the visibility gained SHG to women's work and activities helped to gain a recognition.

One major achievement was the farmers club. The club membership is for husband and wife from a farming family, and not to single farmers. The members meet once every month to share and discuss their experiences and experiments on their current crops. The meeting is held in rotation on different farms - not in house. These were useful in getting the men and women together on crop production technology and interaction was useful in raising the familiarity of technology among men and women together. It helped them understand each other's role and intricacies of managing them. The concept did take roots in only one village out of seven in which it was launched.

MSSM is carrying out an experiment, titled "Aamhi Sudharu Aamuche Gaon" with six Gram Panchayats in District Jalna during the last six months. These Gram Panchayats were identified with the help of District Administration, based on information on the reputation of functionaries and Gram Sevaks/Secretaries. Systematic training and capacity building work has been initiated, with respect to two Gram Panchayats, towards enabling the community and their institutions to vision, plan and mobilize activities for sustainable development of the village as a whole. Regular follow up is being done in form of village visits and organising small activities in these villages. While it is too early to assess the impact of these measures, some visible changes were noticed during the short period of three to four months. They were as follows

- The Gram Panchayats members identified several pressing development needs and identified villages outside the Panchayat body to address some of them.
- The women members have started attending Gram Panchayat Meetings.
- There were a couple of Gram Sabhas actually conducted in these villages; these were attended by a large number of villagers.
- The youth have undertaken several small activities like cleaning of surrounding, awareness on health and sanitation, education, etc.

The proposed project plans to extend these efforts to about 25 villages in the coming three years, and in the process, evolve a pragmatic methodology of capacity building of local communities towards self-governance. Most of these villages would be from Jalna district and a few villages from district Aurangabad would also be included in this project.

Brief Description of the proposed project

The projects aim at capacity building of village communities and their institutions for local self-governance and effective management of State programmes for development of their villages. This would entail enabling and equipping the men and women in these villages to facilitate:

- Democratic functioning of the village institutions, by making the Gram Panchayat and Gram Sabhas more people-oriented and participatory.
- Liaison of men and women with government agencies and other institutions, which would enhance the participation of local communities in development programmes of the State.
- Create a climate of equitable (or at least balanced) development through State schemes.
- Empowering women and bringing them into mainstream of village functioning.
- Orientation of government functionaries, so that they are accessible, helpful and positive.

A multi-pronged strategy is being developed towards achievement of the above objectives. It would comprise of :

- Identification of potential leadership in the villages and government officials and functionaries with positive attitude.
- Orientation of the members of village institutions (mainly Gram Panchayats, Co-operative Societies, other formal and informal committees) along with the potential leadership.
- Coordination with district authorities and panchayat members for their support.
- Involvement of government functionaries and their sensitization and orientation.
- Awareness and skill (managerial) building of the men and women in the villages on development policies and programmes, as also project management skills.
- Skill building through hands-on exercises in participatory problem analysis and project planning - carried out in the villages with real needs and problems.

- Special focus on women and adolescent girls for confidence building and personality development.
- Continuous follow up for landholding and on-the-spot guidance in the fields.

Following chapter is the case study on Adgaon project which is one of the unique projects of MSSM.

References

Alka Srivastava : Non-Governmental Organizations and Rural Development : Social Action; A Quarterly Review of Social Trends; A Social Action Trust Publication, Vol 49. January-March 1999. p. 27.

Ibid, p. 31.

AVARD - Association of Voluntary Agencies in Rural Development, Source - Internet.

Dr. Mrs. Sudhatai Kaldate; Social Change in Marathwada; Development of Marathwada - A Perspective; Publication, Swami Ramanand Teerth Research Institute, Aurangabad, 1999, p. IX.

Dr. Kurulkar R.P.; Economic Development of Marathwada; Development of Marathwada - A Perspective; Publication, Swami Ramanand Teerth, 1999, p. 38.

S.T. Kachwe, G.K. Sangle and V.K. Patil; Opportunities for Marginal and Landless Labourers in Drought Area of Marathwada; Rural India, January 1981.

Manavlok; A Report; April 1982 to December 1987 and Interview with Dr. Lohiya, Chairman.

Nirman, Interview, Shri Usmaan Beg - Incharge.

Janarth : A Report, 1987-1997 and Interview with Shri Koleshwar.

Abhinav Vikas Sanstha; Interview, Mr. Sirsat - Project Incharge.

Grasp; Journal, Professionals for Alleviation of Rural Poverty, Partners in Development, Promoting Sustainable Development Through Collaborative Action, and Interview with Shri Rajindra Singh, former Director.

Institute of Health Management, Pachod (IHMP) and Interview with Dr. Dayalchand, Director.

Dilasa, Range of Activities - Report.

Brief Note on Comprehensive Watershed Development Programme, Adgaon (KD), District, Aurangabad - Implemented by MSSM with People's Participation.

Memorandum of Association of Marathwada Sheti Sahayya Mandal.

Concept Paper : Empowerment and Capacity Building of Communities and Institutions, including PRI's, for Local Self-Governance in Districts - Aurangabad and Jalna, January 2002.

Yugandhar Mandavkar - Information on Marathwada Sheti Sahayya Mandal, December 1997.

Yugandhar Mandavkar, Case Study on Gender and Watershed Management' - Synopsis, Paper Presented for Gender and Technology Workshop organised by Anthra at Hyderabad on July 27-30, 2001.

Concept Paper : Empowerment and Capacity Building of Communities and Institutions, including PRI's, for Local Self-Governance (Aamhi Sudharu Aamuche Gaon in Districts) - Aurangabad and Jalna, January 2002.

CHAPTER - 4

Adgaon Project

Introduction

This project study was undertaken because this work is pioneering and unique in the region and the State. It is one of the most successful programme designed for drought-prone areas in the country. Adgaon has been the drought-prone and most backward village of Marathwada which is now developed into a prosperous and self-sustained village. This project has been a role model in the watershed development and people's participation in Maharashtra and the country. Effects of the Adgaon project can be felt in all the spheres of rural development. viz., people's participation, changes in cropping pattern and land pattern, income generation, underground water level and irrigation, social infrastructure, productivity, job opportunities, health and education, especially in the girls. The project has also stressed on gender-equity and has designed and implemented various women programmes. Through these programmes the women of Adgaon have become socio-economically and politically empowered section of the society. One of the major component of the rural development is animal husbandry and dairy farming, which has proven to be the most lucrative venture of the village today. Afforestation is another important feature of the Adgaon project which led to change in cropping patterns and helping the ecological balance of the area.

In this chapter a detailed study of the project is carried out, this project is popularly known as Adgaon project.The project is a study of village-level integrated micro-watershed development, in a drought-prone area, located on the deccan plateau, and carried out by a regional voluntary agency-Marathwada Sheti Sahayya Mandal (MSSM), with the technical and financial support from various international agencies, national agencies and agencies in Maharashtra of Maharashtra government. MSSM is a voluntary agency working in the central Maharashtra for rural development,in general and watershed in particular. It was started as a public trust to assist small and marginal farmers of the region to improve their living conditions through increased agricultural production and productivity, MSSM undertook several development activities in the land and water management sectors.

Location of the Project

The project is located in the village Adgoan, taluka and district Aurangabad of Maharashtra State. The village is situated between 19^0 47′ and 19^0 48.5′ N latitudes and 75^0 34.5′ and 75^0 35.5 E longitudes. It is located at the northeastern side of Aurangabad City, at a distance of about 30 kms, and is approachable by the Aurangbad-Beed state highway up to 20 kms, and then by an all-weathered road.

Pre-project Conditions

The physical and socio-economic conditions that prevailed in the village, were typically that of a village located in a drought prone area of this region, mostly dependent on rain-fed agriculture and drained by a primary stream (stream having its own origin in same area).

Physical Condition

Topography and drainage of the village landscape is characterized by the existence of undulating topography in the northern parts, and gently sloping flat agricultural farms in the south, with low hillocks (knolls) covering the north, north eastern and north western parts of village area. In the extreme north, the land isvery steeply sloping towards the south, covered with totally barren rocks. The record elevation is different from the highest to lowest point in the area is 45 m, over a distance of about 4.5 kms. A stream locally called Jeevan Ganga, originating from the northern hillocks is the main drain in the project area. This stream flows south, and bifurcates the village area almost into to equal halves - eastern and western. About 11 smaller streams, which drain the area, join Jeevan Ganga at various place as shown in map 4.1. The drainage pattern of the study area is a typical dendritic pattern, in which a number of smaller streams join the major streams, almost at right angles.

Geological Features

The project is located in the deccan plateau of basaltic terrain. The area is underlain by a geological succession of weathered basalt, factored rocks and then hard black basalt. This is followed by soft vesicular basalt and hard basalt. A number of basaltic flows could be identified in the existing well sections, characterized by the presence of a red bole, a red clayey type formation. The weathered and fractured basalt extends up to a depth of 20 to 25 m and forms the only unconfined aquifer in the area. Out of 153 wells in the village, 69 were functional. The average depth of the wells was about 20 m. The soils on the basis of physical characteristics like colour, structure and depth were classified into four groups as shown in Table 6.

Map 4.1

ADGAON PROJECT AREA : SOIL TYPES
GOLATGAON
EKLAHARE
KARAVALA
INDEX
ROAD
STREAM
CART TRACK
INHABITED AREA
SOIL THICKNESS > 50 cm
SOIL THICKNESS 25 to 50 cm
SOIL THICKNESS 7.5 to 25 cm
SOIL THICKNESS < 7.5 cm
SCALE 1: 16000
HUSENPUR
DHARAKWADI
TAKALIMALI
N
GHAREGAON
SOURCE : MSSM REPORT (1989)

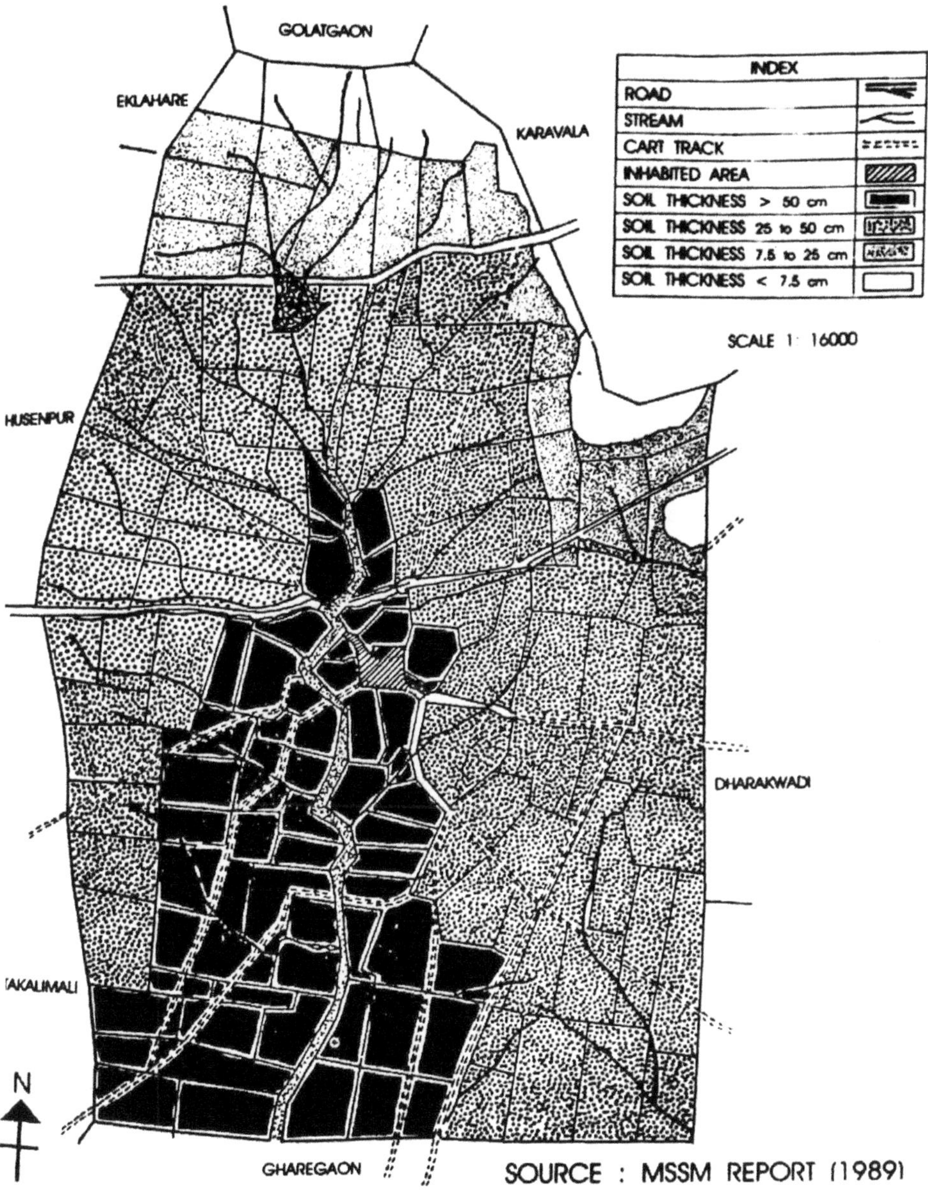

Table 6 : Distribution of lands under various soil types

Depth of Soil	*Type of Soil*	*Slope*	*Condition*	*Area ha¼*	*%*
	Soil	Percent	erosion¼		
Below 7.5 cms	Murum	3 to 8	Very high	28	2
7.5 to 25 cms	Brownish	1 to 3	High	40	4
25 to 50 cms	Brownish black	1 to 3	Medium	580	55
Above 50 cms	Black cotton	1 to 2	Medium	400	39

Source : Modified from MSSM Report 1989, integrated Land and Water Management Programme, Adgaon Khurd, Maharashtra.

The northern parts of the project were characterised by steep slopes and gullies with exposed rock formation. The basic reason for the formation of gullies was soil erosion by the high velocity run-off water. The absence of any vegetative cover or trees added to the acceleration of the soil erosion. The absence of vegetative cover was also another reason for the decrease in the rate of in situ weathering of the basic rock, thereby rendering the soil in the northern parts, infertile.

Rainfall and Climate

The average annual rainfall calculated from the recorded rainfalls over the period 1901 to 1960, was 725.8 mm. With an average of 45.7 annual rainy days. (The nearest rain gauge station was at Aurangabad city, about 30 kms west of the project area). The area receives most of its annual precipitation from southwest monsoon. The rainy season in the project area normally commenced in the first week of June, and was over by end of September. The rainiest month was July with an average precipitation of 173.2 mm. The normal intensity of rainfall in 24 hours was about 75 mm.

The project area, though located far away from the sea, had got fair climatic conditions, having well-defined seasons. June to September was normally monsoon season followed by winter. The minimum mean temperature during winter is 8.4°C. The winter extended up to the end of February, followed by summer, which sets in during March, and lasts till the first week of June. The maximum mean temperature during summer was 42.2°C. The relative humidity data indicated fairly dry weather throughout the year, except the four months of monsoon.

Socio-Economic Conditions

According to the 1981 census, the total population of the village was 1372, including 420 male adults, 423 female adults, 267 male children and 262 female children. Out of a total of 316 households, 275 were cultivator families. The average land holding per household worked out to 3.32 ha, and was fairly high. However, distribution of land amongst the households was quite inequitable, as described in the Table 7.

Table 7 : Land-holding Pattern

Land holding in ha.	*No. of Families*	*%*	*Cumulative holding in ha.*	*Total land*	*%*	*Comul-ative*
Landless	41	13.0	13.0	00.0	00.0	00.0
Below 01	42	13.3	26.3	28.6	3.5	3.5
01-02	84	26.6	52.9	129.6	15.7	19.2
02-03	47	14.8	67.7	120.4	14.6	33.8
03-04	50	15.8	83.5	180.2	21.8	55.6
04-05	6	1.9	85.4	27.6	3.3	58.9
05-06	20	6.3	91.7	112.8	13.7	72.6
06-08	18	5.7	97.4	129.6	15.7	88.3
08-10	5	1.6	99.0	46.8	5.7	94.0
Above 10	3	1.0	100.0	50.0	6.0	100.0
Total	**316**			**825.6**		

Source : Modified from MSSM Report 1989. Integrated Land and Water Management Programme, Adgaon khurd, Maharashtra.

As can be seen from these figures, 85% landholders possessed only about 59% of the land.

Cultural Features

The religions and castes found in the village were Marathas, Brahmins, Sutar, Muslims and other backward classes. People of Adgaon are very religious. The high-class people being more in number the standard of living was very traditional and these traditions and religious customs were very important for the localities. A number of social-deeds took place in the village e.g. Child marriage, discrimination, inequality, superstitions etc. in the name of tradition. People here celebrated festivals like Ganesh Chaturthi, Diwali, Gokulasthami etc.

Education

The level of illiteracy was high in Adgaon. The importance of education and the awareness had not reached the village. Except the school of management and the Zilla Parishad Primary School started lately, there was no other facility for education. For the secondary schooling the children had to go 5 km to the village named Pimpri or 12 km to the village Karmad. This caused only few male children to learn, because sending a girl child even to the local primary school was a rarity.

Health

Due to the above stated illiteracy and superstitions, villagers had to face number of health problems. The other major reason for ill health was the large-scale migration, due to which the earning person of the family left their homes inculcating alcoholism and other bad habits. This caused decline in the health of the people. Thirdly there was no availability of clean drinking water, which gave rise to various contagious and water etc. borne diseases like malaria, flu, jaundice and others like plague, chicken pox etc.

The fourth cause of ill health was that the Primary Health Center was 5 km away at Pimpri and the doctor was hardly available when needed. There were infant deaths due to malnutrition. Due to the lack of the health facilities in the village, women had to face many problems, at the time of the delivery.

As farming was the main occupation of the people, health and well being of the livestock was not satisfactory because there was no veterinary hospital in the village.

Transport and Communication

Upto 8 km there was no facility of transport and communication. This made the people of Adgaon unaware of the happenings in other places. There was no source of information reaching the village easily. This caused problem and delay in accessing medical facilities, no school going, problems relating to the marketing of the products was also felt. All these set backs caused poor standard of living and development of the village.

Women

Women in Adgaon were not different from 70% of the worlds poor population of 3 billions. The socio-economic setup and general environment in the village was not favourable to women. Women faced several problems in the society, mainly regarding education, early marriages, responsibility of stability in the family organisation. Most of the children got primary education, and many went to secondary school 5 kms away from the village. Although facilities were available in the village itself girl children were rarely sent to school. Literacy and education among women was hence very low compared to the men. Girl child was regarded as a great burden on the family, resulting in early marriages. Though the household management was responsibility of the women, she had no say in the financial affairs of the family. The male member of the family took most of the decisions. Women in the village had to bear injustices like co-wife system, desertion, many children to bear, no permission for family planning and mental harassment. Women were rarely seen taking part in village affairs, their only public activity and participation was seen in some religious festivals a few times in the year. As exposure to the outside world is limited for rural women. it is same with the women in Adgaon. Some women in the village did not even visit the weekly market in the vicinity. This lack of exposure also made them more dependent on the men.

Political features

The political features of the village were of general type. Due to the greater number of high-class families, the political bent was towards them. The women had no role or were apathetic to political shifts and thus remained far away from the polity training and leadership training and their fundamental right to democracy.

The involvement during general election was only 20% among the total population of the village. The Adgaon Gram Panchayat was established in 1954, which could hardly make the income of Rs.400 from the local taxes and received Rs.1200 as government fund. This restricted the development of the village in all the directions. Political, social, cultural, economical, etc.

Administrative Features

The political power was with the superior classes, all the administrative facilities were limited to this class. Large number of people stayed ignorant of the various schemes and facilities of government. Being a rural region, people were not politically aware of their rights and also lacked good leadership due to which the administration was not attentive in fulfilling their duties. Administration means it's a duty, a service to people, but it is looked upon as only a salaried job. There was no urge or interest in the development of the region and the welfare programs in the region. All this made the overall development; health, agriculture, education, welfare and development schemes remain in the hands of a limited group of people and were never known to the general public. Also the stringent policies of the government does not allow the administrative machinery to implement and follow program in the rural areas efficiently, which is necessary for the planning of development policies at state and national level.

Land Use Pattern

The village land utilised for various purposes was as presented in Table 4.6. As there was no land reserved under forest, the cultivable area per household, of 3.19 ha, was quite high, while the 83.85 ha of irrigated land was very low in terms of its per capital availability at 0.0026 ha per household.

Cropping Pattern

The major crops in the village were Kharif Jowar and Bajra; and Rabi Jowar, Cotton. Tur (Pigeon Peas), Mung (Green Gram) etc. These were mostly dry crops or rain-fed crops. All the irrigation facilities were available only from dug wells. These were utilised for horticultural plantations. This was mainly sweet-lime with about 24,000 trees. Cropping pattern and crop production for the year 1983-84 roughly estimated by MSSM, was as described in Table 12.

Animal Husbandry and Dairy Activities

Livestock in the project area mainly consisted of 76 cows, 490 bulls and 9 buffaloes, while some families also had a few goats and sheep. In the year 1983-84, there were 40 milch cows and the average daily milk production was about 100 litres. Availability of green fodder was totally lacking, and whatever area under grass cover existed in the village pastureland during the monsoon season, and for one or two months thereafter, was not enough and of little nutritive value. Stock of high yielding, crossbred cows was totally lacking in the project area.

Village Artisans and Other Service Sector Activities

Except for a few service-type units like carpentry, tailoring, flour mill, transport and tractor rentals facility owned by some families from the project area itself, villagers were dependent on nearby villages for all the other agro facilities. This indicated that the situation of the project area was so poor, that it was not able to sustain the units, providing even the most basic services.

Employment and Income

Employment opportunities available in the village during the pre-project period were highly inadequate. About 70% of the total families used to depend on Employment Guarantee Scheme (EGS), 10% of families from this village had migrated temporarily to other villages, and only 20% of the cultivator families were fully employed on their own farms. In the 1983-84, 140 families (60%) out of 231 families surveyed by MSSM were actually below the poverty line, while the average annual household income for the village was Rs.4838, just marginally above the poverty line. Almost 63% of the total income earned was through working either as agriculture labourers, or on works under EGS. Income derived from cultivation on own farms was merely 22%. All this indicated the poor conditions in which the people of the project area were living. Land-holding and occupation-wise distribution of income is presented in the Table 17.

General Conditions and Social Infrastructure

Adgaon was considered the most backward village in the region, so much so, that villagers from the surrounding villages were not ready to give their daughters in marriage to its residents. Nearly 80% of the families from this village were affected by alcoholism. Most of the men - young and old - used to waste most of their time in gossip, or playing cards. There was a lack of capable local leadership, and a sense of concerted community action for overall development. The village settlement was composed of low-grade houses, with thatched roofs and walls and floorings made up of mud. Layout of the settlement was poorly planned, with very narrow lanes. Only 10% of the houses had electrification, and none had their own latrines. For 8 - 10 months in a year, the village was dependent on Aurangabad Zilla Parishad for supply of drinking water by tankers. The cattle sheds were part of the housing premises resulting in filthy conditions, in which members of family and cattle had to live cheek by jowl. There were four Gobar Gas plant in the entire village. However, there was slight improvement in the environmental sanitation due to establishment of several soak pits in the settlement.

The village had a Gram Panchayat office housed in a small, temporarily built room. Tax recovery was very poor. Although the village had its own school, up to seventh standard, it lacked many necessary facilities, like medical dispensary, veterinary aid centre, post office, State transport bus stand, HYV seed and fertilizer distribution centre and crop storage (Godown) facility etc, for which it had to depend upon neighbouring villages.

Genesis of the Project

Barrister Jawahar Gandhi a dealer of tractors and other agricultural implements was instrumental in establishing Marathwada Sheti Sahaya Mandal (MSSM), Jalna registered under Bombay Public Trust Act, (1950) and Bombay Societies Act, 1860 as early as February 1969, and worked as its Honorary Secretary. He was in constant search of a permanent solution, which would mitigate the effects of drought in Maharashtra, and particularly in Marathwada. Earlier, during 1970s, he was successful in providing assistance to farmers, particularly for digging wells, with the help of sophisticated machinery procured by MSSM. However, knowing that whatever he had done for the people so far was not a permanent solution, he was not satisfied with his work, and continued working until he made a big breakthrough at Dev Pimpalgaon village in Jalna District, in 1979.

At this village, where the MSSM constructed its first check dam on a stream, an open meeting of villagers from the surrounding area was conducted, in order to make them aware of the benefits of soil and water conservation and discuss their problems. It was here that he was invited by a group of farmers from Adgaon khurd, to come down to their village and work with them for soil and water conservation. It was during the same time that he found an able associate like Shri Vijay Borade, a graduate agriculturist and social worker, who was later appointed as Trustee of the MSSM, and was instrumental in formulation and successful implementation of the Watershed Development Project in Adgaon.

The MSSM started working with the people of Adgaon during early 1980s. The first few years were utilised for assessing the situation in the Project area and understanding the problems and needs of the people there. By the year 1983-84, a project proposal envisaging inputs to be made by various agencies like Government departments, external agencies and local community, was prepared and sent to Swiss Development Corporation for obtaining funds. Technical experts, from GIT-I unit of Action for Food (AFPRO) , located at Ahmednagar, were instrumental in conducting contour survey at Adgaon and preparing several detailed maps that were incorporated into the project proposal. Meanwhile, various Government agencies and prominent villagers from the project area were approached by MSSM, to help coordinate inputs in the project area in favour of an integrated village level micro-watershed development approach. This was against the traditional approach of making haphazard, uncoordinated inputs dispersed in the larger region and finally not yielding the desired outcome. It was due to various factors like continuous persuasion by the villagers under the leadership of Shri Vijay Borade, overall political tilt in favour of comprehensive Watershed development, and the proactive attitude of government officials at Aurangabad during the period, that government agencies and people modified their plans and made inputs in the project area according to the plans formulated by the MSSM.

Adgaon project of MSSM is basically a watershed development project alongwith all other aspects of rural development. It was at Adgaon that a new model of watershed management was evolved during there last two decades. In 1984, the village had a population of 1,371 (316 families), who were leading a subsistence living, because of poor land productivity and near absence of farm or non-farm employment. The village used to face acute scarcity after monsoon and used to be supplied drinking water by tanker from November onwards. There were 153 wells, of which only 69 were functioning. The average depth of wells was 20 m and water table used to lower to an average 19 m below ground level during summer. All these factors contributed to people's poverty and hardship. Almost all families used to seek wage employment on public works, available in neighbouring villages, from November onwards.

MSSM had been organising an annual get-together of farmers from the region at village Dev-Pimpalgaon in district Jalna, where a micro-watershed was developed using basic soil and water conservation measures. Almost all villagers of Adgaon visited this project during 1981-83. They felt that such approach is urgently needed to overcome the perpetual drought situation in their village and sought guidance and assistance from MSSM. The dialogue went on for some time, and finally in 1983, MSSM started small experiments in Adgaon, with technical inputs from AFPRO, Agricultural University, ICRISAT, WALMI, and the State Government. By mid-1984, a plan was prepared jointly by MSSM and the people, with technical support of AFPRO. It was implemented during 1984-88 with support from SDC (then Swiss Development Corporation) under AFPRO-SDC Phase-II programme.

The project was aimed at drought-proofing with a goal to improve the physical quality of life of people by increasing their income through appropriate land and water management on watershed basis, and to provide assured employment opportunities. Accordingly, activities like afforestation; contour bunding; plantation of trees and grasses to meet food, fodder and firewood needs of the community; rainwater harvesting; etc. were undertaken. The project also undertook the promotion of appropriate water use systems like drip irrigation, non-conventional energy sources like bio-gas and smokeless *chulha*; water supply, sanitation; and vocational training for landless and women.

Philosophy of Watershed

A watershed is basically any area of land that sheds or drains water into a puddle, pond, ditch, lake or river. The water that is being drained is water that neither evaporated nor soaked into the soil. Therefore, the lands are from which the water drains into a given section of a river is the "watershed" relative to that section of the river.

In a simple language we can say that watershed embraces all the land and water areas, which contribute run off to a common point. The term watershed' refers to the separating one drainage basin from another. So the term is used synonymously with catchment and drainage basin. In short watershed is a hydrologic unit which produces water as an end product by the interaction of precipitation and watershed factors. The quantity and quality of water produced by the watershed are an index of the amount and intensity of precipitation and the nature of watershed management[4].

Watershed management is concerned with the rational use of the soil and water resources in the watershed, so as to at least sustain and it is possible that the productivity level through a combination of engineering, agronomy and afforestation measures.

The rags to riches story of Adgaon village stems from its ecological development through watershed management. Adgaon village falls in the acutely drought prone area and was facing the problem of soil erosion arising out of surface run off in times of scanty rains. The control of runoff and soil losses was the basic problem reducing the productivity of rainfed agricultural lands. Soil erosion not only affects the productivity but also reduces the effective life of their water resources. Problems such as sedimentation] tree decline, soil acidity and salinity, declining soil structure and fertility affect their total catchment. It was observed that an estimated three tons of soil were lost per year. Policies were to restrain this high rate of erosion and drought. It is very pertinent to note that it takes hundred of year to build an inch of it by nature. So one of the major components had been soil and moisture conservation and giving vegetative cover to the soil by way of developing grasslands. In fact soil conservation has been an important activity under the agricultural development programmes from mid-fifties. However, most of the lands in drought prone areas remained prone to erosion.

The major objects in the hilly catchment area of Adgaon were :

- To reduce the incidence of drought and the intensity thereof.
- To increase infiltration water, in the soil.
- To control excess runoff damaging the soils.
- To harvest and utilise runoff for recharge ground water resource.
- To undertake suitable land management programmes to make production base safe for a high level of sustainable production.

All this was possible in Adgaon only because it was a need-based village with the active involvement of the local population.

Project Plan of Adgaon

The main objective set by MSSM to be achieved through this project was to enhance the standard of living through increased productivity, and assured employment by improving management of land and water. The village Adgaon was selected as an area of operation for two-feasibility considerations- social and technical.

It was thought that it would be easier to work with a village community, already tied by social bonds, willing to work in collaboration with MSSM and technically, this village had well demarcated mini-watershed area of only about 1000 ha, which was easier to handle for a NGO like MSSM, with limited capacity of implementing such intensive programmes.

The strategy adopted for achieving the objective, included a set of programmes to be implemented under the project, as mentioned below:

- Water harvesting and surface water development through a series of check weirs and earthen dams;
- Soil conservation through series of gully plugs;

- Aforestation and agroforestry to meet the requirement of fodder, fuel and fruit as well as for soil conservation;
- Dry Land Treatment and Management through contour bunding, farm frains and strip cultivation to accelerate the soil building process;
- Water Conservation through efficient irrigation methods like drip irrigation, appropriate cropping systems and dry horticulture;
- Development of non-conventional energy resources, to meet the energy requirements of the village;
- Rural water supply and environmental Sanitation; and
- Extension Services for Farmers and Vocational Training for landless women.

The first five Programmes were envisaged to deal with the development, conservation and improved utilisation of two basic resources land and water. The next two were felt to be essential for fulfilling basic needs of the people like fuel and water, apart from its indirect contribution towards soil and water conservation, and the last was justified as being essential for transfer of watershed management technology for ensuring people's participation, and for increasing productivity of the human resources.

Details of specifications of various inputs planned to be made under these programmes were as follows:

Water Resources Development

Before 1984, a percolation tank was constructed by the government in the northern parts of the area with a storage capacity of 0.43-mcm. However, as the catchment area was limited, only 0.11-mcm could be stored during the normal rainfall year. Also, as the percolation tank was located in the uplands of the watershed, where a few of irrigation wells were located, much benefit could not be derived from the percolation tank. In the proposed plan 13 storage structures, seven on the main stream - Jeevan Ganga - and six on the subsidiary streams had been planned. All these thirteen structures, including the percolation tank, were estimated to store about 0.5-mcm. The storage structures were planned in such a way that the majority of the existing wells located in all parts of the watershed would get the benefit of recharge. These storages were to function as recharge bodies well meant.

Depending on the social and ground conditions, two types of structures had been chosen namely, overflow and non-overflow structures. The overflow structures were chosen to be constructed in the middle and lower parts of watershed, where land slopes were gentle and the adjoining lands were deep vertisols or fertile lands. In such places, the submergence of land for storage and loss of fertile land posed a big problem. Hence, these had to be low structures storing as much water as could be accommodated in the streambed itself. Again, on the basis of cost considerations two types of overflow weirs were considered: First one being two-meter high dams, with masonry type construction and overflow regulated by gravity, and the second type being more than two meter height dams, with combination of earthen and masonry type construction, known as reinforced earthen dams, with overflow regulated by gravitational forces. Earthen dams with separate section for overflow/outlet were planned to be constructed at places, where dam site was located in the marginal and or degraded lands, and the submergence was estimated to be of moderate size of up to 2.0 ha only, and was agreeable to the people.

A total of 9 masonry weirs and four earthen bunds/dams were planned to be constructed under the project. The total cost estimated to be incurred on this programme including cost already incurred on percolation tank was Rs.15,78,000.

Soil Conservation through Gully Plugging

To arrest the soil erosion as well as to check the flow of run-off water in lands with steep slope, a series of gully plugs were planned to be constructed. A total of 56 gully plugs made up of earthen dams, having capacity to store about 0.168 mm of water were proposed to be constructed. The total cost estimated for this purpose was Rs.3,18,000.

Dry land Treatment and Management

As mentioned earlier, the project area was classified into 4 soil types. Different set of treatment works were proposed to be implemented in different soil types areas, such as :

Deep Vertisols (Black Cotton Soils with soil thickness more than 50 cms): Proposed treatment package for this soil type, covering 400 ha (roughly 40% of the geographical area of the village) consisted of land shaping works mainly using ICRISAT's method of laying broad-based ridges and furrows along with farm drains with the help of agricart, and then carrying out agricultural operations along with the contour. This type of treatment to land was proposed to be given routinely by farmers while performing regular agricultural operations, hence no additional cost was to be incurred for this purpose.

Light Shallow Soils : Land with soil depth ranging from 25 to 50 cm and with slope of 1 to 3 % was classified under this category. The land with this soil type covering about 55% of geographical area of Adgaon was proposed to be treated with the package - consisting of contour bunding and trenching across the slopes, broad ridges and furrows with ties, farm drains and farm ponds. Spacing between the bunds was decided to be of 6 to 10 m, depending upon land slopes and as well as on prevailing agricultural practices. Trees for protection of bunds were proposed to be planted on bunds; whereas strip cropping of rain-fed crops was proposed to be promoted. The cost per hectare, excluding the cost of saplings, worked out to be Rs.910.

Light Very Shallow Soils (Brownish Coarse) : Land with Soil thickness between 7.5 to 25 cm and slope between 3 to 5% covering about 40 ha (4% of geographical area) was classified under this soil type category. The treatment package proposed to be rendered to this soil type was the same as that for the light shallow soil type, except that it was proposed to promote development of grasslands in between the bunds instead of cultivation of rain-fed crops.

Hillocks and Rocky Land (Murrum) : Land with rocky exposure or soil thickness merely of less than 7.5 cm, covering 28 ha (2% of geographical area of the village) was categorised under this soil type. Treatment works proposed for it were intensive afforestation with trenching and contour bunding. A land treatment and management programme was planned for about 600 ha of land at an estimated cost of Rs.6,28,370.

Afforestation

This programme consisted of two schemes namely, the promotion of Agroforestry on private farms, and tree plantation on mass scale on rocky and murrum soil. The selection of trees was to be made on the basis of factors like their fuel, fodder and fruit value; their contribution to soil conservation and soil-building processes; resistance to drought; deep-rootedness and rate of growth. The species proposed to be selected were Ber (Berries); Limba (Neem); Bhendi; Kawat (Wood Apple); Chinch (Tamarind); Neelgiri (Eucalyptus); Sewga (Drum stick); Biba (Marking nut); Vasa (Bamboo); Sitaphal (Custard Apple); Dalimb (Pomegranate); Jambul (Black Plum); Amba (mango); Rita (Soap tree); Subabul (Lucena); Tur (Pigeon pea-Perennial Variety) and Avala (Gooseberries). The grass species selected for promotion on inferior soil type lands were Hemata and Madras Anjan. A nursery was proposed to be established and maintained by MSSM, in order to provide saplings of required quality and quantity, in time. The estimated cost for this programme was Rs,5,56,000.

Water Conservation - Drip Irrigation System

It was observed that in the existing system of flood irrigation, a large amount of water is wasted by way of evaporation and transpiration. So, the drip irrigation system was proposed to be introduced at least for orchards (horticultural plantations) covering about 100 ha.

Further, it was proposed to establish the drip irrigation system in at least two plots of one hectares each for using these as demonstration farms; and it was envisaged that the people would themselves take up this irrigation system in their orchards of the size of 100 ha, in due and it was estimated that the total cost for introducing this system into larger area of 100 ha, would be about Rs.12,78,000.

The remaining three programmes namely, development of non-conventional energy sources, water supply and sanitation and extension and vocational training are not discussed here in detail. Their specifications are briefly presented in Tables 8 and 9. The set of inputs proposed to be supplied under these programmes were not directly related to soil and water conservation in the village, however some of these were expected to be helpful for facilitating this process. For example the input called Establishment of 60 gobar gas plants was supposed to meet the fuel requirement of about 30% of the households, and hence was expected to be helpful for protection of tree plantations and in turn, for the soil conservation in the village. Similarly, extension and vocational training to be provided under the last programme, was expected to generate skilled human resources, capable of carrying out independently, the maintenance of physical assets created under the first five programmes and for adoption of new agricultural practices in the village.

Table 10 : Programme-wise physical assets created, contribution made by various agencies and benefits generated through the project.

Programme	Physical Assets	Cost Contribution			Total	Benefits
		MSSM,	Govt.,	People		
Water resource	08 check dams	548,00	-	-	60,000	0.965 m. cum water storage level in about 100 wells and providing protective irrigation to Kharif and regular irrigation to Rabi crops in 500 ha.
development	01 check dam		60,000	-		
Soil	04 Gully Plugs	20,000	-	-		0.558 m. cum water impounded
conservation	27 Gully Plugs		642,000		662000	conserving soil eroded from large area of about 300 400 ha.
Dry Land	230 ha treated	336000	-	-	1063000	40 ha of community pastureland and about 550 ha of agriculture land with conserved soil and enhanced fertility.
Treatment	270 ha treated	-	444,000	283,000		
	77 ha treated					
Afforestation						
a) Waste land	125 ha waste land plantation	242,000	-	N.A		Thousands of trees conserving soil & shelding fodder, fuel, .
plantation	22 ha waste land plantation					
		-	-	N.A		About 80 ha of horticulture plants with plantations of sweet lime fruit-33 ha, Mango-41 ha, sapota-18 ha.
b) Agroforestry	1.3 ha mango plantation	-	2000	-		Pomegranate-7.5 ha, Custard
	41.3 ha Horticulture scheme	-	120000	-		Apple-6 ha., Guava-1 ha.
	40.0 ha Horticulture by people	-	-	N.A		
Water Conservation	12 Drip irrigation	N.A.	N.A.	N.A.	N.A.	16 ha of orchards covered, water saved for additional irrigation.

Programme	Physical Assets	Cost Contribution			Total	Benefits
		MSSM,	Govt.,	People		
Non-Conventional Energy Sources	72 Bio-gas plants constructed] 16 functional¼	N.A.	N.A	N.A.	N.A.	Fuel provided to few families.
Water Supply & Sanitation	01 Overhead tank built on hilltop, supplyingv water to community stand posts.	N.A.	N.A.	N.A.	N.A.	Assured drinking water supply through stand posts.
	16 latrines	N.A.	N.A.	N.A.	N.A.	Hygienic practice established Prevention from water borne diseases.,
	130 smokeless stoves	N.A.	N.A.	N.A.	N.A.	Women & children protected from smoke.
	50% of settlement area provided with open lines drainage system	N.A.	N.A.	N.A.	N.A.	Environmental sanitation ensured, prevention from Malaria.
	80% houses electrified	N.A.	N.A.	N.A.	N.A.	Better lighting of houses at night time
	Pucca houses & settle-ment with better layout	N.A.	N.A.	N.A.	N.A.	Better dwelling units and better surroundings.
Extension Services & Vocational Training	Several Educational Tours, Training Classes Seminars and Group meetings conducted.	N.A.	N.A.	N.A.	N.A.	More skilled and better organised human resources.
01 to 08 TOTAL		1146000	1551000	NA	2697000	

N.A. = Data not available.
Source : Data collection from MSSM.

It is observed that foreign funding agency, different agencies of Government of Maharashtra - like Agriculture and Minor Irrigation Departments and Directorate of Social Forestry - and people of the village, have contributed to the various works carried out under the project. It was difficult to keep account of funds contributed by people for various works. The contribution from various State agencies, was essentially for construction of a check dam and 27 gully plugs, land treatment on 350 ha, plantation of trees on 40 ha community pastureland and horticulture scheme on 40 ha, whereas, Swiss Development Corporation and MSSM had contributed for construction of 8 check dams, 4 gully plugs, land treatment on 230 ha and tree plantation on 125 ha wasteland etc.

Apart from this, various works not directly related to soil and water conservation were carried out in the village. These works included establishment of 72 bio-gas plants, 16 latrines, 130 smokeless chullahs, open (lined) drainage system, and electric connections to 70% of the houses, and rehabilitation of people in a settlement with pucca houses and better layout, and construction of an overhead tank on a hill-top supplying water to community standposts in the village. The various agencies like MSSM, Government of Maharashtra and beneficiaries themselves contributed to various extent for these works.

The numerous additional works carried out in this village not planned earlier are as follows:

- Establishment of public library, which was gifted a set of books on agricultural farming;
- A T.V. Set gifted to Gram Panchayat;
- Construction of a bridge near main settlement;
- Balwadi Building;
- Establishment of meteorological mini-centre with equipment like Rain Gauge, and Eater Evaporation Rate Measurement Instruments by Meteorological Department of Government of Maharashtra;
- Distribution of fertilizers and seeds to 85 beneficiaries under Land Scheme of Marathwada Agricultural University;
- Construction of additional structure at the outlets of gully plugs;
- Raising the height of few check Dams and the desiltation and deepening of the stream-bed at check dams; and
- Grafting on various horticultural plantations.

It is observed that MSSM executives with active involvement of officials from Soil Conservation Department had played lead role in coordinating the implementation of these works.

Leadership

As mentioned earlier, two executives of MSSM, Barrister Jawahar Gandhi and Shri Vijay Borade, were mainly instrumental in planning and execution of various works under this project. Shri Gandhi being keenly interested in finding a solution to drought in Maharashtra, and Shri Borade being graduate in arts and a progressive farmer himself, with high aptitude for social work, made a capable team of leaders for this purpose. Because of their capabilities, they were able to understand quickly the specifications of relatively simple technology and skills of managing group dynamics in a village community, required for micro watershed development. Shri Borade, due to his rural background, was well conversant with regional culture and language was aware of agricultural and socio-economic conditions prevalent in a typical village of this region and the people's felt needs. So he was able to work effectively for organising people, maintaining liaison with district level and local officials of Government, while Shri Gandhi possessed all the skills required for managing office records, drafting of proposals, monitoring of implementation process and conducting concurrent evaluation of the project. They were constantly improving their knowledge and capabilities by consulting technocrats and bureaucrats from Departments of Soil Conservation (Agriculture), Minor Irrigation, Social Forestry, and Ground Water Survey and Development Authority, Zilla Parishad Aurangabad and other agencies of Rural Development Departments of Governments of Maharashtra.

They had consulted technical experts from Action for Food Production (AFPRO) for conducting various surveys and formulation of project proposal, ICRISAT for Land Treatment Works and Agricultural Universities like Rahuri and Marathwada Agricultural Universities for bringing about changes in agricultural practices.

They were not striving for monetary gains or any political advantage. They had the motto of social service, were public oriented, and had maintained continuous contact with people for at lease 5/6 years through group meetings and sometimes, even by staying in the village. Due to their charisma, and listening capacity, they were able to resolve any dispute arising due to submergence of private land under reservoirs and / or its acquisition for other public works. This enabled them to successfully reallot land to various affected people, all cases being resolved at village level, without referring them to taluka and District level government organisations. Due to these capabilities they could act as catalysts in the village, to bring about a complex process like watershed development, requiring multi-agency co-ordination.

People's Participation

As indicated earlier, people were consulted and kept informed during the planning phase of the project, about the various proposed works to be carried out in the village. In fact, they helped various survey teams by providing them access to their land, and to inform about their assets, treating them to a cup of tea and food. During the initial stages of implementation, an executive committee comprising one prominent person from each ward/group from the village was appointed by the MSSM with the help of general body meeting held at this village. This committee was to liaise between MSSM/Government functionaries and villagers, to ensure community participation in all the public works to be carried out. The information provided in Table No. 10 indicates various contributions made by the people towards the implementation of the project. It is observed that, besides community lands, several hectares of private land were also treated under the project. The guide bunds were constructed by people offering voluntary labour- shramdan. They also planted several thousand trees on their own land, and developed community pastureland by this form of collective voluntary effort.

In order to protect community and private grasslands and tree plantations, people were convinced by executive committee members and MSSM, to practice stall-feeding for their cattle stock. Some women from households possessing more sheep stock, were initially reluctant to practice this, and were taken on a study tour to Rahuri, where the long-term benefits of tree plantations were demonstrated to them. Thus they were convinced to sell off their sheep stock, in favour of protection of grassland and tree plantations. In order to render support to landless families affected due to ban on open-field animal grazing, the MSSM and executive committee procured cross-bred cows under Government Schemes, and allowed them to take home one headload of grass and other minor produce from community pastures and other public lands. A social penalty of Rs.25 per grazing animal per day was imposed, against open field grazing.

People were consulted frequently for deciding the location of sites for check dams and gully plugs. Some of them parted with their land, which got submerged under water reservoirs, and some even shared their land with others, who were affected very badly by this. They had contributed with shramdan for the foundation work of several check dams.

Executive committee members supervised these construction works under the MSSM direction. Due to submergence of a road connecting the main settlement to other villages, the people were in trouble, and collectively approached the Zilla Parishad Aurangabad to get sanction for a culvert to be built on this spot. This work was completed partially by shramdan, in the record time of just 21 days. Similarly, due to submergence of various foot and cart tracks parallel to the stream, the owners of nearby land were approached to allow the villages to use some narrow patches of their land for public thoroughfares. The owners of such land were convinced by the MSSM and executive committee of villagers, of the necessity, and they readily agreed to do so.

Apart from all this, people were required to bring about desirable changes in cropping pattern, and were expected to make full use of enhanced sources or irrigation, by laying down micro-irrigation systems in their own farms. Several families invested their share for procuring pump sets, electric connections and laying down irrigation channels in their farms. They also diversified their cropping pattern - crop varieties yielding fuel and fodder were grown on a large scale in the village, for the first time. People adhered to the norms laid down by the executive committee and MSSM, and agreed to the community prohibition of cultivation of any water-intensive crops like sugarcane and bananas in the village. In order to save water, some farmers brought their land under drip-irrigation (16 ha 1991).

Some of the people even participated in the vocational training, imparted to them for maintenance of physical assets generated under this project. They even shared the cost of educational tours conducted to various demonstration sites and construction of a balwadi building. They had improved their housing stock and established gobar gas plants and smokeless chullahs by sharing the cost. By 1988-89, due to the success of the Project, Adgaon had gained a lot of publicity, and the number of visitors including Government officers. Officers from NGOs, Journalists, social science researchers and various groups of farmers, started visiting the project and it became necessary to attend to these visitors and brief them about the achievements of the project. Some of the visitors coming through the MSSM office at Aurangabad, were being attended to by the MSSM's functionaries themselves, whereas visitors directly landing up at the village, were either attended to by the village-level officer appointed by the MSSM, or by the group of people oriented for this purpose. In this village people very enthusiastically took the visitors around the village and showed them the various works carried out there. Apart from providing information to the visitors, this group also arranged for the visitors' meals.

Impact of the Project

Under the impact of the project, Adgaon Village has been completely transformed into a drought-proof and prosperous village. The various changes physical, and socio-economic that have occurred in the village are as mentioned below:

Major Economical : Due to soil and water conservation activities the productivity levels have gone up to the income of agricultural produce in 1991-92 was to the tune of 70 lacs at as compared to 11 lacs in 1984-85. Major increase is mainly due to the watershed development.

The availability of ample fodder and water, however, goats from the village were sold and the number of cross breed animals is increased. Similarly the investment in agricultural machines and implements has gone up.

In an economic impact survey, it was revealed that out of 97 families indicated increase of income levels from Rs.500 to 50,000 per annum. Even the wages of landless labourers have been substantially increased. Some 21 families from nearby villages have migrated back to Adgaon with the increase in employment period and wages. No work force is now available for EGS works, which was major source of survival before the land treatment. The group of marginal and small farmers had experienced major prosperous changes with watershed management and change in crop pattern. Instead of traditional crops they have harvested the crop of mausambi (sweetlime). Area under cash crop has increased so rapidly that it has covered two third of the total area.

Social Impact

The transformation of the village is not only economic but also the villagers now have greater unity and are committed for the improvement of community life. They have become more vocal and confident. Before the implementation of the project, they were unwilling to accept new things, but now they are constantly in search of new agricultural techniques.

The villagers first took the decision to ban the goats from the village to protect the vegetative cover. Till today no stray cattle is allowed to graze freely and illegal cutting of trees is completely prohibited.

In the spirit of community participation that the villagers had collected village funds for further development and maintenance like deepening of streambed.

The gram sabha of the village has become the highest democratic body where the programme to be implemented has to be approved by the unanimous decision. With the effective working of gram sabha, no elections of society or gram-phanchayat were held since 1986. Once the decision of gram sabha always remained final without any dispute. No case was reported to the police station as the decision of gram sabha was socially accepted. Many small and seemingly insignificant houses have been converted into pucca houses. To make the village totally self sufficient in all its needs was MSSM's dream, which was fulfilled at the larger extent.

The recent survey on the impact of watershed on women's life exhibited interesting results. With the watershed development, it was found that women were over burdened. However, they did not take the increased work as a burden but with joy and fulfillment.

Technical Impact

Various organisations like the Swiss Development Corporation, the AFPRO, the WALMI and the irrigation department conducted different technical surveys on the impact of Adgaon model.

The groundwater recharge analysis with a quantitative impact assessment by SDC found drastic change in groundwater recharge. The checkdams show their principal tillty in raising the water table early in the monsoon period by allowing infiltration of the early runoff waters.

It was also proved that the contour terracing has a strong impact on the recharge of rainwater. With the effective watershed management, the recharged volume increased by 85 percent in recharged areas and 30 percent in rainfed areas and over the entire watershed, the recharge coefficient increased by 40 percent from 15.4 percent to a maximum of 21.5 percent.

The empirical analysis of the impact of comprehensive watershed development, conducted by WALMI reached to the conclusion that the involvement of voluntary organisation MSSM in Adgaon could achieve peoples active involvement in watershed development work. The study also probed deep into the impact analysis of landless labourers with socio-psychological aspects. It was found that there was positive significant impact of watershed development on landless labourers as compared to landowners, probably because the migration was totally stoped. In all the surveys it was admitted that 160 wells have been recharged and 500 ha acres of land is covered under irrigation. Spill erosion has been controlled. Moreover, horticultural crop pattern was accepted by farmers. The demand of labour has substantially gone up. Significantly eight landless farmers have purchased lands.

Land Use Pattern and Cropping Pattern.

As a result of watershed development works, area under irrigation has increased from 83 ha in 1984 to 500 ha including about 140 ha of land under perennial irrigation in 1990-91. The availability of cultivable area has increased from 790 ha in 1983-84 to 1005 ha in 1990-91. Details of the changes brought out in Land Use Pattern are as presented in the Table No. 11.

Table 11 : Changes in the Land Use Pattern

Land Use	*1983-84*		*1990-91*	
	Area in ha.	*%*	*Area in ha.*	*%*
Cultivated Area under Irrigation	83.85	7.99	500.00	47.65
Rainfed Cultivated area	706.41	67.32	505.00	48.12
Cultivated Area	790.26	75.31	1005.00	95.77
Waste Land	217.00	20.68	0.00	0.00
Gairan	38.00	3.62	0.00	0.00
Settlement	4.11	0.38	8.11	0.77
Sub merged area	0.00	0.00	36.26	3.46
Cultivable Waste	259.11	24.69	44.37	4.23
Total Geographical Area	1049.37	100.00	1049.37	100.00

Source: Modified from MSSM Report 1989, Integrated Land and Water Management Programme: Adgaon Khurd, Maharashtra.

This has enabled the rain-fed crop cultivates of this village to bring their land under various cash crops and under the horticultural plantations. Apart from increase in the cultivable area and irrigation facilities, enhancement of extension services and seed, fertilizer and pesticides distribution services also have contributed to a great extent for achieving agricultural development in the village.

Changes in Cropping Pattern.

The changes observed in the cropping pattern are as presented in the Table No. 12.

Table 12 : Changes in Cropping Pattern

	Crop	*1983-84*			*1991-92*		
		Area (ha)	*Production*	*Yield (qntls/ha)*	*Area (ha)*	*Production*	*Yield (qntls/ha)*
I	**Kharif**						
1.	Bajra (Pearl Millet)	500	600	1.20	800	2000	2.50
2.	Tur (Pigeon Pea)	150	50	0.33	300	200	0.66
3.	Mung (Green Gram)	10	10	1.00	---	---	---
4.	Kapus (Cotton)	60	00	0.00	100	300	3.00
5.	Holga & Matki	200	50	0.25	300	200	0.66
6.	Groundnut	---	---	---	10	25	2.50
7.	Sunflower (comm.)	---	---	---	300	650	2.17

	Crop	1983-84			1991-92		
		Area (ha)	Production	Yield (qntls/ha)	Area (ha)	Production	Yield (qntls/ha)
8.	Sunflower (seeds)	---	---	---	100	200	2.00
9.	Vegetable	---	---	---	100	N.A.	N.A.
10.	Green Fodder	---	---	---	125	N.A.	N.A.
	TOTAL	920			2135		
II	**Rabi**						
1.	Jowar	N.A.	00*	0.00*	500	800	1.60
2.	Kardi	N.A.	00*	0.00*	N.A.	50	N.A.
3.	Harbara	N.A.	00*	0.00*	100	40	0.40
4.	Wheat	---	---	---	200	400	2.00
5.	Sunflower (comm.)	---	---	---	200	600	3.00
6.	Sunflower (seeds)	---	---	---	50	80	1.60
7.	Vegetable	---	---	---	75	N.A.	N.A.
8.	Green Fodder	---	---	---	50	N.A.	N.A.
9.	Mosambi	27	00*	0.00*	200	N.A.	N.A.
	TOTAL	N.A.			1375		

* In 1983-84, due to scanty rainfall, though the indicated crops were sown, the production was practically nil.

N.A = Data not available

— = Not applicable, as these crops were not cultivated.

Source : Data Collected from MSSM.

Animal Husbandry and Dairy Development

Due to watershed development works carried out in the village, availability of nutritious green fodder has increased tremendously. This has enabled the villagers to maintain more livestock and diversify their income earning activities to cover small-scale dairy farming also. (Gross cropped area under green fodder which used to be almost nil has increased to more than 175 ha, including 125 ha in *kharif*, 50 ha in *rabi* and few ha in summer in the year 1991-92).

Some indicators of growth in dairy farming sector in the village are as presented in Table No. 13.

Table 13 : Growth in dairy farming sector

Sr.	*Particulars*	*1983-84*	*1991-92*
1.	No. of local variety cows	40	00
2.	No. of cross-breed	00	335
3.	No. of dairy farmers	40	126
4.	No. Average daily milk collection (in litres)	100	864
5.	No. Selling price for milk] in Rs. (lit.)	2.25	5.50

Source: data collected from MSSM

The total income of the village, only from dairy farming, has increased from about Rs 81,000 in the year 1983-84 to Rs 15,84,000 in the year 1990-91

Village Artisans and Other Services Sector

Growth in the agriculture and animal husbandry sector has enhanced the employment opportunities in all the other sectors of village economy. Units serving the village people for agriculture operations, processing and transport of produce, and for better consumption have flourished, and the growth in the number of service units and the income generated by them, is shown in the Table No. 14.

Table14: Growth in service sector

Sr.	*Service Unit*	*1983-84*		*1990-91*	
		No.	*Transport Vehicles*	*No.*	*Income in Rs.*
1.	Transport Vehicles	01	50,000	05	3,30,000
2.	Tractors	01	3,500	05	1,20,000
3.	Threshers	N.A.	N.A.	05	28,000
4.	Floor Mills	01	4,000	11	25,175
5.	Tailoring	01	3,000	11	40,000
6.	Carpentry	01	4,000	01	14,000
7.	Bullok carts	11	N.A.	81	N.A.

N.A. = data not available
Source: Data collected from MSSM.

Employment and Income

The village economy prior to the implementation of the project had highly inadequate employment opportunities. This had compelled 10% of the villagers to desert the village and stay in other villages, 70% of them had to work either under employment guarantee scheme or as agricultural labourers in other villages. In the village, there were only 20% of them fully employed in their own farms. Now, after the completion of several watershed developments works, the village economy, in terms of employment and income, has witnessed tremendous growth in all its sectors like agriculture, animal husbandry, dairy farming, and other service sectors. The villagers who had deserted Adgaon have returned, and there are about 20 families who have immigrated to this village from resource-poor villages in the surrounding areas, while several people from nearby villages have been regularly visiting Adgaon to work on agricultural farms. The daily wage rates have increased from Rs.10 and Rs.5 per unskilled male and unskilled female labourer respectively, in the year 1983-84, to Rs.30 and Rs.15 in the year 1990-91. At present, nobody from the village is seeking any employment under EGS. Tremendous growth has been witnessed in the incomes from agriculture and subsidiary occupations like animal husbandry, dairy farming, horticulture and Agroforestry. Details of the sector-wise growth in annual income of the village are presented in the Table No. 15.

Table 15 : Sector-wise Growth in Income

Sr.	*Sector*	*1983-84*	*1990-9*	*Growth (No. of times)*
1.	Cultivation	2,47,990	48,89,575	19.71
2.	Subsidiary Occupations	1,60,500	14,52,600	9.05
3.	Service/Labour	7,09,200	11,63,220	1.64
	Total Income	**11,17,690**	**75,05,395**	**6.71**

Source : Data collected from MSSM.

It is noteworthy, that miraculous growth has occurred in total income, and income from cultivation and subsidiary occupations at the rate of 6.71, 19.71 and 9.05 times respectively, in the very short period of only seven years. This is all the more significant, as the income figures are considered for the year 1990-91, which was a scanty rainfall year with only 490 mm of rainfall.

Another favourable change that has occurred, is in the sector-wise proportional contribution to the total annual village income. Earlier, (during 1983-84) income from service or labour, earned mainly through working on EGS works used to be 63% of the total income, whereas now (in 1990-91) the major contributor to total income is crop cultivation activity, at 65%. Details on sector-wise contribution to total village income is as given in the Table No. 16.

Table 16 : Sector-wise Contribution to Income (in percentage)

Sr.	*Sector*	*1983-84*	*1990-91*
1.	Cultivation	22.19	65.15
2.	Subsidiary Occupation	14.36	19.35
3.	Service / Labour	63.45	15.50
	Total Income	**100.00**	**100.00**

Source : Data Collected from MSSM

Details of increase in income for different land holding categories and from different sectors are as given in the Table No. 17.

Figures in parentheses are percentages to concerned totals, whereas figures in additional (second) parentheses under column nos. 6 and 10 are annual incomes per household in Rs.

Source: Data collected from MSSM.

It is significant to note that average annual income per household has increased more than 6 times, i.e. from Rs 4838 in 1983-84 to 32,491 in the year 1990-91.

General Condition and Social Infrastructure

This village which was a drought-prone, resource-poor and disorganized community area, is now successfully transformed into a prosperous, drought-proof and high social development area. The villagers are now no more found gossiping at public places, but are busy instead in learning methods and techniques of increasing the productivity of their of their farmlands, and bring more prosperity to their house holds. Due to implementation of micro-watershed development projects and increased awareness of the people, general conditions in the village have improved dramatically. Growth and development of various social infrastructural facilities are described in the Table No. 18.

It is observed that this project, implemented in a short period of just 5 years, with funds from a foreign agency and Government organization has achieved lot of favorable changes in the village. Substantial increase in irrigation facilities, diversification in crop cultivation and horticulture plantations, increase in crop and livestock yield, generation of adequate income and employment opportunities, are the important outcomes of the project. Changes achieved in social infrastructure are also significant, although equitable distribution of increased irrigation facilities was not deliberately attempted. Nonetheless, the ban imposed on direct lifting of water from reservoirs, fair dispersed distribution of water storing structures all over the village and economical use of water through drip irrigation systems, have to some extent achieved this. The chief merits of this case lie in preplanned and short duration implementation of various works under this project. The proposal of this project, prepared by MSSM with the help of AFPRO, gives a detailed description of various techniques and methods to be adopted for replication of village level micro-watershed development project in other villages of this type. After successful experimentation with this village, MSSM has formulated detailed project proposals for covering some of the villages from surrounding areas, again with the help of funds to be procured from various foreign funding agencies and state government departments.

Table 17 : Size of the Land holding and Source-wise Income.

		1983-84				1990-91			
Cate-gory	*No. of house holds*	*Cultivation*	*Subsidiary Occu-pation*	*Services Labour*	*Total*	*Cultivation*	*Subsidiary Occu-pation*	*Services Labour*	*Total*
1	2	3	4	5	6	7	8	9	10
Land-less	17	Nil	9,000	31,400	40,400	Nil	71,000	76,500	1,47,520
	(07.74)	(00.00)	(22.28)	(77.28)	(100.00)	(00.00)	(48.13)	(51.87)	(100.00)
					(2,376)				(8,678)
00.0-02.5	18	5.975	4,000	35,500	45,475	95,175	51,700	1,04,400	2,51,275
	(07.79)	(13.13)	(08.08)	(78.06)	(100.00)	(37.88)	(20.57)	(41.55)	(100.00)
					(2,526)				(13,960)
02.5-05.0	57	18,990	19,500	2,90,000	3,28,490	7,53,100	3,30,400	4,05,00	14,89,100
	(24.68)	(05.78)	(05.94)	(88.28)	(100.00)	(50.57)	(22.19)	(27.24)	(100.00)
					(5,763)				(26,124)
05.0-10.0	70	65,515	52,500	1,79,700	2,97,715	11,56,500	3,12,500	2,98,900	17,67,900
	(30.30)	(22.01)	(17.63)	(60.36)	(100.00)	(65.42)	(17.67)	(16.91)	(100.00)
					(4,253)				(25,256)
10.0-15.0	35	35,110	13,700	79,600	1,28,410	8,41,200	1,90,500	1,56,800	11,88,500
	(15.150)	(27.34)	(10.67)	(61.99)	(100.00)	(70.78)	(16.03)	(13.19)	(100.00)
					(3,696)				(33,957)
15.0 +	34	1,22,400	61,800	93,000	2,77,200	20,43,600	4,96,500	1,21,000	26,61,100
	(14.72)	(44.16)	(22.29)	(33.55)	(100.00)	(76.79)	(18.66)	(04.55)	(100.00)
					(8,153)				(78,268)
Total	231	2,47,990	1,60,500	7,09,200	11,17,690	48,89,575	14,52,600	11,63,220	75,05,395
	(100.00)	(22.19)	(14.36)	(63.45)	(100.00)	(65.15)	(19.35)	(15.50)	(100.00)
					(4,838)				(32,491)

Table 18 : Growth and development of various infrastructural facilities

1983-84	*1993-1994*
Village settlement crowded, accommodated in an area of 4.11 ha.	Village settlement spread in planned manner into an area of 8.11 ha.
Housing stock included kutcha buildings made up of thatched roofs and walls and flooring made up of mud.	Housing stock improved to include many newly build houses with brickwork and RCC structures.
Settlement provided with Soak pits.	Sock pits replaced by open lined water drainage system laid down in village settlement under JRY scheme
Not a single latrine block in entire village.	Out of the 230 proposed latrines, 16 latrines have been completely builds, and works for remaining are going on.
10% houses were electrified.	80% houses are electrified.
Traditional chullahs were been used.	130 smokeless chullahs are established, in addition 35 solar cookers have been distributed.
Drinking water was being supplied by tankers by Zilla Parishad for atleast 8 months in a year	Piped water scheme established by utilising CAPART funds.
4 Gobar Gas plants were established.	72 Gobar Gas plants constructed, out of which 16 have commenced functioning.
HYV seeds and fertilizers were seldom used, only 10 tonnes of fertilizers were being used for sweet lime plantation.	Use of HYV seeds and chemical fertilizers has drastically increased. About 100 tonnes of fertilizers are being used. A shop storing HYV seed and fertilizers is established in village
Godown for storage of agricultural produce was not available.	Land for a godown building has been acquired and it is proposed to be set up in next one or two years.
Milk collection center was available, but average daily milk collection was 100 litres. Only local variety cows were available.	Average daily milk collection has gone upto 1400 litres in the year 1991-92. Milk producer's society has in its account a deposit of Rs. 75,000/- collected as commission.
Veterinary services were not available.	Govt. and private veterinary doctors make frequent visit.
Primary school upto 7th class.	School has been extended upto 10th class. Proposed hostel block is under construction.
Medical facility was not available.	Govt. and private medical dispensary located in village itself.
70% eligible couples covered under family planning.	100% eligible couples covered under family planning.
State transport bus stop was not available.	State transport bus stop is made available and also in addition private vehicles available for transport.
Gram Panchayt office accommodated in an old and small building. Tax recovery was very poor.	New building constructed. Tax recovery 100%.
Only one community level formal organization i.e. co-operative society was established.	Various local level public organizations have been established like milk producers co-operative society. Educational institute, Lift irrigation co-op. Society, Public library, Gymnasium Balwadi/Anganwadi, Mahila and youth club etc.

Source : Data collected from MSSM.

The remarkable work done by the people of Adgaon is now being seen by the Maharashtra Government as a model for implementation of watershed development programmes. The villagers have dramatically and drastically changed the face of the village within four years through voluntary work.

Before 1986, the village was facing acute problem of drinking water. Rainfall figures indicate that Adgaon has suffered from drought for 28 out of the last 78 years, which shows that one out of three years has been a drought year. Most of the villagers had to work on employment guarantee scheme (EGS) for their survival. Literally 90 per cent village was moving outside the area in search of the work. Some of the small farmers were working on each other's land due to shortage of labour. Uncontrolled grazing and cutting of trees had left the hillocks surrounding devoid of vegetation. There were only 8450 trees in 1984 in the total 1063 hectares area of the village. Now Adgaon is a green island as the grasslands are developed on shallow, lightsoil, hillocks and rocky lands with intensive afforestation by trenching and bunding. The species were so selected that they could yield timber, bio mass fruits, firewood. At present there are 50,000 horticultural plants. About 3.75 lacs trees have been planted on the wasteland while social forestry department have planted 70,000 plants in 38 hectares of Gairan land and a lot remaining to be done.

People of Adgaon are still recollecting the days when they were insulted by adjoining villagers by refusing to hold any matrimonial relations with Adgaon youths. Such type of poverty and humiliation they suffered for many years only because the village was dry with the scanty rainfall of 500 millimeters a year. In 1987 the area received only 300 mm of rainwater by which 152 wells were dried up and that there was no water found even 80 meter below the ground. From 1972 drought to 1986, the government had to supply drinking water to Adgaon by tankers.

The milk production was merely 100 litres in 1983, which has now crossed the limit of 1000 ltrs. There were the days when people migrated from the village for livelihood. After the watershed development in Adgaon, all migrated people returned home, moreover 21 labourer families have come to Adgaon from other villages. Before the land treatment, there were no effective industries in the village. Now the villagers purchased matadors, thresher machines, tractors. The village is now having single phase industrial floor mill & tailoring units. In 1983 the village income from all resources was only 11,17,690 while in 1990-91 it was estimated around 75 lacs.

Adgaon Watershed Development Project was launched in 1984, after about three years of preparations to organise and orient the community on the principles and practices of watershed development. It was the first ever project of its kind, in terms of scale, implementation strategy and community involvement. At that time, it was believed, in absence of any field-data, that watershed development, being a land-based programme, would increase the disparities among people by benefiting the "haves" and excluding the "have-nots" from any development benefits - an equity issue.

Most of the soil and water conservation activities in Adgaon Project were completed by mid-1987 and the benefits started flowing in from the very next crop season. The changes were observed regularly and the results were discussed among development practitioners with a view to analyse and understand the processes. A detailed socio-economic impact assessment study was conducted in 1991-92. The results showed that the project had achieved considerable equity across classes, with the landless and small and marginal farmers benefiting far more proportionately that the erstwhile "better-off" farmers.

Attempts were also made to study equity on social aspects in the project. Prominent among them was the Gender Analysis Study conducted in 1993-94. This study highlighted the disparities among men and women in terms of benefits, workload, technology and options generated by the project.

An impact assessment study was undertaken by MSSM in 1991-92 to understand the changes in socio-economic aspects of the lives of people in Adgaon. The study was designed with the help and guidance of practitioners, social scientists and academics from other VOs, Universities, AFPRO and WALMI, Aurangabad. Data was collected on pertinent social aspects and economic activities of the people through household survey, group discussions, observations and direct measurement. Process documentation, regular recording of events and physical phenomena MSSM and were useful in drawing certain conclusions during the study.

The village had only one percolation tank constructed by the government during 1972 drought. No work was taken up in the village on the grounds of technical feasibility. MSSM undertook the work of check-weirs and nalla plugs. Thirty-two nalla bunds and gully plugs and eight masonry plugs were constructed in series on small and big streams, with collective water storage of nearly 17 Mcft. In the meantime, the state agricultural department came forward to help the MSSM and constructed one check-dam, 27 gully plugs with the dry land treatment on 270 ha and land grading on 77 ha.

Due to soil conservation by contour bunding and plantation of trees and grasses on bunds, the soil erosion was controlled - the fertile top-soil was retained in the land and the falling rainwater was soaked into the ground. This ensured better moisture availability during the crop growth season in dryland areas, reducing uncertainty with the farm production on account of unpredictable rainfall.

Due to combined effect of soil conservation and rainwater harvesting structures, the recharge to groundwater increased significantly. All the 153 wells in Adgaon have been functioning since 1989-90 to provide protective irrigation for two seasons. The irrigated area increased from 63 ha in 1984 to over 500 ha in 1990.

The land use and cropping pattern had a sea change during this period. The farmers, including the marginal and small farmers, had adopted newer technology and crops, and through experience, have developed the attitude for technological improvement. Instead of traditional crops, farmers preferred horticultural crops like sweet lime and vegetable. Area under cash crop, mainly vegetables, has increased so rapidly to cover two third of the total area.

The improvement in soil and water situation resulted in overall growth in agricultural and allied sectors, increasing the family income many-folds. Extension of cropping season to over 200 days in a year provided ample employment opportunities to all, resulting in all-round prosperity.

Livestock Development : The milk collection at the cooperative dairy society increased from merely 100 litres per day in 1984 to about 1,000 litres. The estimated value of milk production was Rs. 28 lacs during the year 1991, which was a drought year.

Income levels : The total family income from various sources like agriculture, livestock, service sector, etc has shown tremendous increase, from less than a thousand rupees per year to about 15,000 per year, particularly the landless and the small and marginal farmers.

The effect on the landless was prominent. The income of landless families increased from Rs. 332 to Rs. 15,000 per annum. Similarly, the wages of farm labourers have substantially increased. It is interested to know that the rise in income of the landless and the small and marginal farmers had been more that proportionate as compared to the large farmers. Many landless families have purchased land and their economic condition is better than some so called "better-off" farmers.

People have given a priority to health and education. Educating a girl child is every family's pride. Today, it is nearly impossible to find a girl, who has not studied up to secondary level. The villagers purchased a piece of land and constructed two rooms for the school by local contribution. The involvement of village leaders in the functioning of school was helpful in maintaining the educational standard.

People of Adgaon still recollect the days, when they were insulted by neighbouring villages by refusing to hold any matrimonial relations with Adgaon youth. Such poverty and humiliation they had to suffer for many years, only because the village was dry and poor. Now that the villagers have become affluent, the marriages in Adgaon are sometimes extravagant and the dowry rates have also reportedly increased.

Gender equity : Gender analysis study of the project was carried out by Dr.Uma Ramaswamy and Ms. Bhanumathy Vasudevan and supported by SDC in 1993-94 to understand the impact of the watershed development on women in Adgaon. This study brought out some striking disparities among men and women with respect to the benefits.

The programme was focused on family as a unit and not on men and women separately. Although the benefits reach the family as whole, the men were found to play an important role in deciding on the use of the benefits, mainly financial and social benefits. It was observed that the workload of both men and women has increased, but the women had to share a larger proportion of it. While the workload of men-folk could be made easy with the availability of tools and technology at his disposal, the women were completely denied the access to technological advancement in the course of development.

The equity in Adgaon was mostly the result of natural economic processes. MSSM's efforts in this direction, were, in that sense, limited to bringing all people together to plan, implement and monitor a resource-based programme and to constantly cautioning them of possible pitfalls.

The benefits to small and marginal farmers accrued mainly because they started cultivating their own lands, before they took work on a large farmer" land as labourer. Most operations on their own land were done in time, whereas the larger farmers, who did not have adequate family labour to cultivate his entire holding, had to suffer losses due to delays. The landless mostly benefited because of liquid cash they could earn. With increase in wage rates and availability of work round the year, their income increased significantly. They could buy lands out of the income they received in wages. Similar trends were noticed in Jadgaon project, undertaken by MSSM during 1992-97.

Gender equity in Adgaon had remained a desirable, but distant, goal. Subsequent to the study, some efforts were made in Adgaon to address women's issues. In future projects of MSSM, gender intervention is planned from the very beginning, so that the women develop a sense of confidence, pride and identity. Opportunities provided to women to play lead role, even in the smallest of social or economic activity in the village, in building their capacity.

Advocacy Role: MSSM took conscious efforts, albeit not very systematic, to promote the watershed approach on the backdrop of Adgaon during 1988-93. It was backed by several studies conducted by researchers and academics and a continuous monitoring of happenings on key parameters of socio-economic development. Appropriate to the development thinking of late 80's and early 90's, the focus was on technology and integrating it in the social fabric of rural areas.

Adgaon project was successful in incorporating certain appropriate technologies in State policy and programme, such as contour bunding in place of broad-based furrows in shallow soils, masonry structures, and spillway treatment of earthen structures. The community approach was useful in formulation of pedagogy for Indo-German Watershed Development Programme (IGWDP). The gender aspects in Adgaon project, and subsequent experimentation in Jadgaon project, helped in including a specific component in the IGWDP then.

Adgaon project demonstrated a viable and replicable approach to resource development on watershed basis. Adgaon project, a national example of integrated watershed development, was helpful to development planners and practitioners to formulate policies and programmes for equitable development for poverty alleviation and drought proofing on sustainable basis.

Women Empowerment

The achievements not withstanding, Adgaon project had shortcoming in two areas, viz, the project could not evolve, as learning, a well defined the strategy for community organisation, and secondly, women's involvement. While the efforts for community organisation were mainly adhoc and often limited to locally solving problems as they arose, the aspect of women's involvement was largely ignored in planning and implementation of the project.

With respect to women involvement, attempts were made only at a later stage in the project. Initially, women were not separately approached in the development planning, primarily because it was believed that women and children are integral part of the family institution, which gets represented by the farmer (or head of household) in public affairs. This aspect was looked into later, through discussions with several development workers and through formal reviews; but it did not help alter the programme strategy. It was only after the project was over, some meaningful analysis could be done.

Perspective on Women's Involvement in the Project

The above experiences and observations and that of others in the field highlighted the following aspects in relation to women's role and position in society. This understanding, although based on prevailing situation in rural areas, can be broadly generalised for the entire society and stated as premises in the following paragraphs.

Although most work in agriculture and almost all-domestic chores are done by women, society usually avoided recognising the importance of women's role in these spheres of life. Society, by and large, discouraged the involvement of women in social activities and public life, probably because women are generally believed to "lack competence", and hence, not given any opportunity to prove their capabilities.

The above factors, along with the prevailing systems of social sanctions, deprived women of any opportunities of confidence building and induced inhibitions. Conventional development pattern generally failed to consider women's role in planning and operationalisation of development efforts.

Object of Intervention for Women

Based on the above premises, MSSM attempted for "Improvement in status of women in the social, economic, opinion leadership and attitudinal spheres of the community" as a long term goal, using the medium of "Involvement of women in watershed development". The strategy towards attaining the goal could have the following components;

- Inculcate a habit of coming together to discuss various aspects of women's life in family and within the community. This is for women's organisation into mahila mandals.
- Facilitate, through above process, an analysis of and an ability to analyse women's role in situation related to drinking water, kitchen sanitation, health and nutrition, etc. to increase a critical awareness on such "seemingly unimportant" activities.
- Promote and facilitate economic independence through savings and thrift and through increased employment opportunities.
- Identify existing skills and technologies possessed by women and build upon the same. This would entail skill development training and upgradation of existing technologies in farm and domestic sector.
- Identify leadership qualities in women and encourage directive channelisation of such abilities.
- Facilitate "recognition" to women and women's activities within the family and in the community. This would also necessitate gender sensitization training and orientation of men and family.

Recent experiments of MSSM

MSSM decided to adopt the above strategy in a phased manner. Presently, attempts are being made to organise and activate mahila mandals and to initiate the process of getting "recognition" to women and women's activities through awareness and exposure programmes. Attempts are also under way to identify specific intervention needs in the areas of income generation, skill-development and technology-upgradation, formal education, and sanitation. Subsequently, as the second phase, appropriate methods and approaches to address the intervention needs are planned to be designed. In third phase, experiences are proposed to be analysed with a view to extend the process and programme to other areas and VOs.

Achievements: For little over a year, MSSM has been able to organise mahila mandals, and through their regular meetings, promote the concept of self-help and income-generation. As a starting activity, savings and thrift groups are formed. As yet they have not been able to identify appropriate ways of multiplying their savings. The mahila mandals are planning to link their savings activity with income-generation and are in the process of identifying and experimenting with economic activities like food processing.

Issues concerning household chores are ordinarily being discussed at present, primarily in the areas of nutrition, health and sanitation. Training programmes were arranged to familiarise women with simple food-processing techniques, which focused more on nutrition than income generation. Increasing the awareness of women on concepts and components of watershed development is a pre-requisite for their active involvement in the process, awareness camps and guided exposure visits have been organised for the purpose. Simultaneously, attempts are made to impart technical skills required in short-run, e.g., nursery raising, spillway construction, masonry, etc.

Difficulties Encountered : Although the response from men and the rest of the village community to the above experiments had been warm, they are looking at it with some skepticism. Some men have expressed the need for such efforts, but they seldom have actively offered any help[7].

Organisationally, MSSM was not fully geared to take on the task because getting sincere lady workers was a major difficulty. The atmosphere in villages of Marathwada is not as yet very conducive for male members of the society to conduct women's meetings. Training facilities available in the vicinity, too, are normally not suited for organising women's programmes. Similarly, women in the villages are already overburdened with their routine activities, and thus, find it difficult to spare much time for mahila mandal work.

The women have generally confined themselves to traditional domains of activities, mainly in the domestic sector. This has helped reinforce their self-image as well as induce a shyness to venture into new areas of public activity. A lack of self-confidence and the fear of unknown prevent women to attempt any move in this direction. Similarly, uncertainty about community's response, especially the male quarters, make them hesitant even to talk about any of their aspirations or intentions.

Opportunities, backed by community will be an assurance for even an indication of acceptance by the community would certainly help women to come forward and start such initiatives, that help them gain confidence to work towards the goal. MSSM's efforts in the direction of facilitating and reinforcing the confidence and positive self-image of women are recent and are yet to offer any clear trends. The attempt is to help women gain skills knowledge and information in the first place and provide opportunities to them through such activities that would help them establish their identity.

Finally, it must be noted that the women in Adgaon started talking about their development and related issue came forth only after their basic needs were met. Experience, thus far, in other villages, too, indicates similar limitations on part of women and communities to venture into empowerment issues in situations of poverty and inadequacy.

During the study many households were visited to collect data and found that the life in that village has been totally changed. Personal observations and interviews of many villagers, Gram Panchayat Sarpanch, Secretary Dairy Co-operative, Chairmans of various Self-Help Groups revealed the following aspects of the Adgaon village society;

- Now no families migrate or go outside the village in search of labour.
- Previously those who were labourers have now got their own land.
- It is difficult to find labourers in the village as now everyone has their own farms and labourers are called from other villages.
- Thus people from other villages are coming to Adgaon in search of labour.
- The average annual income of every farmer has increased nearly to 4-5 lacs.
- There are 25 SHGS formed in which 3 are of men.
- Most of the villagers participated in the various activities done by MSSM, viz., tree plantation, contour bunding, contruction of watershed building of allas etc.
- Today also there is a village level watershed committee which looks after the management and repair of the watershed.
- The persons other than farming occupation, e.g. Lohar (weilder) have increased their income twice.
- Every farmer has his own plantations.
- The rate of literacy has raised to 90% and women are now 50% literate.

Dairy is one of the distinguishing feature of the village Adgaon. The villagers and the secretary of the dairy give the credit to MSSM and Shri Vijay Anna Borade. During the project implementation in 1982 with due guidance and leadership of Vijay Anna the first dairy (Gyanraj Milk Production Co-operative Society) was formed. This dairy has 70 members and is one of the standard series category A' in the State. Annual income of this dairy is 50 lacs. In 1994 the Sant Tukaram Co-operating Society was formed, it has 90 members and in 2001 Saraswati Women's Co-operative Society was formed, which constitutes of 27 members so today Adgaon has three dairies, daily 1700 litres of milk it collected. The 10 days income from this collection varies from 2 to 2.5 lacs including the payment of farmers and dairy commission.

The highlighting features of the Gyanraj dairy is that they saved 5 lacs rupees from the commission received by the government, by which they will be constructing a new building and other accessories for dairy. It also gives Rs. 10,000 loan to their members on their own capacity. With the introduction of dairy, there are some changes in the life of the peoples they are -

- Dairy has brought a joint-venture business for the farmers in the village.
- People got labour by bringing fodder for cattle.
- This has benefited the women more and at the same time overburdened them with additional responsibilities.
- Dairy increased the livestock and fertile land due to cowdung fertilizer.

Thus the introduction of MSSM's watershed project has completely changed the life of people in their every aspect - socio-economic, physical, cultural etc. Till date there is a worker and some women social worker of MSSM, who stay in Adgaon to look after the problems of the people, help them in sorting out difficulties in getting loans, addresses the women, self-help groups and also utilizes villagers help in nearby villages, where similar watershed projects of MSSM are being implemented.

References

Summary of the paper prepared and presented by MSSM at the Impact Assessment Workshop organised by AFPRO for the parterns of AFPRO-EZE-CA pacakge-IV at Aurangabad on August 4-8, 1997.

Ramesh Vaswani., Micro-Watershed Development., Three success Stories from Maharashtra., Published by Yeshwantrao Chavan Academy of Development Administration 1995, p. 50.

Ibid., P. 52

Brief Note., Comprehensive Watershed Development Programme. Adgaon] KD . District : Aurangabad., Implemented by MSSM with peoples participation.

Ramesh Vaswani., Micro-Watershed Development., Three success Stories from Maharashtra., Published by Yeshwantrao Chavan Academy of Development Administration 1995, P. 56.

Yugandhar Mandavkar., Equity in Watershed Development., A Case Study on Adgaon Project., Presented at the Workshop on Participatory Watershed Development organised by Swees Agency for Development & Co-operation's] SDC Project Support Centre for the Partners of Indo-Swees Participatory Watershed Development Project - Karnataka] ISPWD-K at National Institute of Agricultural Extension Management] MANAGE ., Hyderabad on September 2-4, 1998.

Yugandhar Mandavkar., Watershed Development : Shifts in Involving Women., Paper presented at the Workshop on "Women's Involvement in Watershed Development" organised by People's Action for Watershed Development Initiatives] PAWDI project at Jaipur on July 29 - August 1, 1996.

Interviews with the people of Adgaon.

CHAPTER-5

Peoples' Participation In Rural Development through Voluntary Agencies

Introduction

This chapter studies the democratic theory, concept, meaning and need of people's participation. The important factors of participation people's are also discussed in the chapter. The role of NGO's in brining about participation of people in the rural development process is highlighted with the observations made and the voluntary agencies studied by the researcher.

The base of democracy is the active and efficient participation of the people in the affairs of government. Democracy demands that public administration must be based on the public opinion, consent and support. The actions of public agencies and officials are expected to reflect the aspirations, interests, demands of the support potentials of the public they serve and direct. They also need to command active support and cooperation of the people. Democratic responsiveness by administrative officials and responsible citizen's involvement in administration are considered as pre-conditions for an effective administrative process in the modern polity. Two of the major criteria suggested for determining whether an administration is democratic are : " 1) It must be open in the sense of having wide contacts with the people and 2) it must be controlled not only by an official hierarchy but also by public opinion."

In order to fulfil the above mentioned conditions of effective and democratic administration the government has to maintain wider and regular contact with the people through consultations and exchange of views on vital issues of decision making and ensure citizen's active participation in administration[4]. It becomes necessary on the part of the government to devise some formal institutional innovations like Joint Consultative Committees, Advisory Councils or Vigilance Committees. These committees enable the government to be aware of the state of mind of the governed[5] and help the administration to mobilize public support and participation in its activities. This has rightly been described as active citizenship' upon which the success of administration lies.

In short, it may be said, that the projects spring up as soon as the funds and personnel are available by the donor agencies and the projects cease to omit once the agencies withdraw their support. If development is one-time procedure, then the question of continuity of efforts or sustainability would not arise. But it is not so. Hence, there is a need to think of means of generating self-sustaining efforts so that the projects once initiated would go along and produce continuous benefits.

Increasingly, sustainable development stress that development must be participatory and must involve local people in decision-making process that affects their lives. The people, for whom the development programmes are meant, must take part in planning and execution for the practical reason, that development efforts which do not involve local people / groups often fail. In this context, peoples / community participation in the development process becomes essential.

Earlier, it was thought that technological transfer and capital transfer would bring about development. But the thoughts of the programme implementers failed to recognise that the development must come from within and not brought out by outsiders. It failed to recognize that equity is a necessary condition of development and the development will not be sustainable if rich become richer and poor become more poorer.

After studying the development programme of developing countries, the need for local participation is stressed. The consensus is that if development is to be sustainable, it must be participatory and community based. The development initiatives should be based on the needs identified by local people and involve them in the designing and implementation of the projects using principle and techniques suited to local condition. This process would include all sections of the community, specially women.

Democratic Theory and Participation

Democracy is cherished by people all over the world. All those who are in power loudly talk about peoples mandate behind their actions and activities. Over thousand of years, People have discussed and debated about the meaning of democracy and how best to initiate and sustain a democratic system of governance. One workable definition is Democracy is that system of community government in which by and large, the members of a community participate, or may participate, directly or indirectly, in the making of decisions which affect them all.' In other words, a system can be called democratic when there is scope for participation in decisions, directly or otherwise.

Political scientists perceived participation as an ingredient of every polity. Those who fail to participate, whether out of neglect or exclusion, are likely to enjoy less power than others. Although not all who participate possess effective power but those who do not participate cannot exercise or share power. The above observations imply that the right to participate is an essential element of democratic government. From the ancient times to present, the political philosophers have extolled popular participation as a source of vitality and creative energy as a defense against tyranny, and as a means of enacting the collective wisdom.

By involving the many in the affairs of State, participation should promote stability and order; and by giving everyone the opportunity to express his own interests, it should secure the greatest good for the greatest number. The community should gain, further more, by drawing upon the talents and skills of the largest number of people. Some philosophers have claimed that participation benefits the participants as well as the larger community. It enables men by giving them a sense of their own dignity and value, alerts both rulers and ruled to their duties and responsibilities and broadens political understanding.

However, there are some scholars who even consider that participation can be used to bypass the elected members of parliament and can be highly undemocratic. Devolution of important decisions to local bodies may mean handing power to members of the local power elite who grind the faces of the poor. Central decision making often provides safeguards for the interests of poor.

When conventional methods of development fail to yield any significant results, there is need for some rethinking about the socio-economic instructions that link resources, people and government. These institutions are the backbone of economic models of development. In a system of planed development, conventional wisdom generally prompts the introduction of government as an alternative agent to people' in the process of taking and implementing decisions. While macro-economic decisions, actions and policies regarding flow of resources, financial management, trade prospects, price formulations, etc., are well within the basic realm of the government, its role in bringing about sustainable development at the rural grass-root level is becoming increasingly doubtful. Various alternative formulations of rural development programmes have been attempted. Community development programmes, Panchayati Raj Institutions, Integrated Area and Rural Development Programmes, National Rural Employment Guarantee Programmes, Block Development Models and so on, are instances of different ideological approaches opted for, at different points in time, in the process of rural development aimed at people' at the grass-root level.

Rural development that is undertaken without properly identifying links between various socio-economic factors and agents at the village level can impose contradictions within the system, leading to conflicts and failures, for which the people are generally blamed. Ignoring the economic links between private property resources is not a shortcoming of the people. It is a shortcoming of the planning processes itself. The history of human societies all over the world has witnessed the close link between land, water, forest, minerals and people. Resources are used by the community, with a great deal of interaction between individuals. The interaction is of a socio-economic conformity resulting in the joint determination of the use of, and access to, resources. Allocation of resources based on such a community approach, rather than on an individualized one, adds an element of sustenance to the development process.

Such a development programme, focussed on the village community in particular, will have to take a fresh book at common property resources, people's access to them, both legal and conventional, their links with private resources, and its sustainable aspects. This will have to be done, further, from the viewpoint of the people. This new outlook calls for the emergence of a fresh social institution to be called participation'. Participatory development therefore, forces aiming for sustained development at the village level.

Concept of Participation

The concept of participation is an old one. The ancient Greek Scholar, Aristolte found a clear relationship between the extent of participation and the creation of a good life. According to him, the best State was where there was broad participation with no class dominating the others. His analysis showed some relationship between the participation and development.

According to one school of thought participation is equated with democratic values and believes that it will produce many additional positive results. If people take active part in solving their own problems and meeting their own needs they will acquire the power that is retained by government through default.

Development professionals have advocated a school of thought which urged a humanistic - democratic participative management philosophy as the only practical way to get the beneficiaries committed to the project and to build local capacity.

The concept of participation is often used as a slogan without careful consideration of precisely what it implies. The term has been used by different people to mean a good many different things. Several scholars in the academic as well as administrative world have discussed and defined the concept of participation. Among them are Pateman, Esman, Uphoff, Montgomery, Ramesh Arora and Sastry. Inspite of the enormous importance attached to the concept of participation in recent development literature, there is hardly any consensus as to what it signifies. In different words, the term participation has not been conceptualized clearly and it continues to be vague and imprecise and quite often it is becoming increasingly rhetoric. Most of the above mentioned scholars have raised certain questions regarding participation. First, there is the question of the purpose of participation; the second is what form should participation take? and the third, what is the relationship between participation and democratic institutions ?

Community participation is defined as people acting in groups to influence the direction and outcome of development programmes that affect them. The key aspects of this definition are people acting collectively and influencing the outcomes. Collective action is necessary, as many aspects of watershed development require co-ordination between neighbours and among members of communities. Their influencing outcome is the core of participation. Participation at a basic level is to share and to own. It is a process of mutual adjustment and orientation of behaviour among participants, not one doing what the other wants mutual adjustment requires communication and interdependence as actions of one depend on the actions of another. Therefore, participation is a voluntary process which may take place in several stages beginning with appreciation in which various parties listen to each other, share perceptions, and validate each other's information . Information sharing is followed by negotiations, joint decision-making, and implementation.

Meaning of Participation

The term participation has its origin from Latin word participare' which means taking part' encyclopaedia of psychology describes participation as : 1) "Taking part or involvement in an activity", and 2) "Greater involvement of persons in policy decision which affect them directly." It simply connects self-activity in relation to social and economic situation.

Cohen and Uphoff regarded participation as "generally devoting the involvement of a significant number of persons in situations or actions which enhance their well-being, eg., their income, security or self-esteem". Paul defined participation as "in the content of development, community participation refers to an active process whereby beneficiaries influence the direction and execution of development projects rather then merely receive a share of project benefits.

M. L . Santhanam defined participation as "commitment on the part of the individual towards all forms of actions by which the individual can take part' or play a role' in the operation without being conscious of any socio-economic barriers to achieve certain common goals in a group situation. "Such commitment or involvement would be possible only if he is effectively appraised about the situation so as to enable him to form an attitude based on his own perceptions. This implies that for effective participation, people of all class and caste should be involved in all stages of plan process. To bring about participation in reality the bottom-up' approach should be prevalent than the top-down approach.

After examining the large number of factors carried out in the rural development tasks, Lele concluded that local participation might mean involvement of people in planning including assessment of local needs. Even if the people do not participate they must be well informed about the plans designed for their blocks. Participation in planning and implementation of programmes can develop self-reliance, which is necessary among rural people for accelerated development. But most of the times it is observed that people in poor communities are the recipients of the improvements' which had little connection with their problem.

The ILO points out that participation involved active, collectively organised and continued efforts by the people themselves in setting goals, pooling resources together and taking actions which aim at improving their living conditions.

ACC Task Force on Rural Development views participation in terms of active' and passive' process. It points out that what gives real meaning to popular participation is the collective effort by the people concerned to pool their efforts and whatever other resources they decide to pool together, to attain the objectives they set for themselves.

Understanding the concept of participation, to effectively participate, persons or group have to identify themselves with the movement. This would be possible only when people are involved in need identification, fixing priorities among the needs, and associated in the planning, decision-making, implementing and evaluation process and given an opportunity to create or maintain or manage local organisations. In general it can be said that community participation implies development as a whole, with and by the people, not just for the people.

Participation means different things to different organisations. Baumgartnar suggested that for some, participation is persuading communities to accept whatever is being proposed to them. This is true even if some projects that place high value on participation. At the other extreme, a small minority of implementing organisations views participation as a fundamental right of the communities, and a means to empower them.

John Cohen and Norman Uphoff in their study on rural development participation classified rural development participation in four ways;

- Participation in decision-making
- Participation in implementation of development programmes and projects.
- Participation in sharing the benefits of development; and
- Participation in monitoring and evaluation of development programmes.

The importance of the rural poor participation in planning and implementing efforts to improve the quality of their lives has been affirmed in recent years by extensive research and experience. This participation includes roles for local people and assessing their needs, organizing local development activities, implementing those activities, committing resources to them and assessing their impact. While much has been written about the benefits of participation, it largely boils down to the common sense notion that people tend to support and commit themselves to activities in which they feel some ownership. In turn, such ownership is fostered by real involvement in planning and managing such activities.

Important Factors of Participation

1. CPR (Common Property Resource)

The most observable situation in which participation emerges is the existence of common problems. Drought, flood, soil erosion, epidemics or wars do seem to force people to come together to take decisions regarding common action. The logic of such collective action is extendable to the management of most CPRs.

Government rules and regulation did not yield result initially, as they target individual with lower social responsibility who can violate them with or without government support.

The next alternative is imposition of rules and regulations from outside agencies on the village communities as a whole. The beginnings of the participatory process are found in acceptance of such a regulatory mechanism. The situation has elements of common interest, an essential input into any cooperative effort. At the third stage, self-enforced societal rules such as stall feeding or social fencing are set up and self-policing replaces outside agents; and the institution of regulated individual behaviour and social responsibility with respect to common property resources emerges as a stable one.

The degree to which participation can develop in the context of a particular village economy depends on its socio-cultural and economic structures. One important aspect of this structure is assests distribution. In economics where a large percentage of households have access to private property resources (PPRs) in the form of land or livestock, it is easier to set up rules for the management of common property. This is so, because links with their private property resources can be established either to increase their productivity or shift the economy to a somewhat higher level of development. In both cases, a process of development that is sustainable is initiated. The participatory development process, based on the creation of links between CPRs and PPRs in the region, is found to be stable and sustainable. In that sense, the type and levels of links bring about different degrees of participation. The extent of the stability of the development process is partly a function of the nature of the community assest. If the assest, such as a community forest, is less susceptible to variations in output as between good and bad rainfall year, village incomes resulting from it will be stable. Alternatively, an irrigation tank, which may dry up in a drought year, will not be able to impart stability to incomes generated.

2. Leadership

In all instances of participatory development the existence of leadership seems to be one of the first catalysts, though not the only one. Identification and realization of commonalities, shedding in individualized value systems, recognition of the greatest common good of the greatest number of people, foreseeing the evils of the tragedy of commons, etc., require mass appeal, demonstration, and intervention in present day social life. Leadership can start the process of consciously homogenizing the village economy, which is otherwise permeated by social, economic and caste stratums. Leadership also seems to have played a role in bringing to knowledge of the people the significance of managing CPRs for a sustainable development and the cost of breach and mistrust in social connections, norms and responsibilities. Leadership can either emerge within a society or come from outside, and it may subsequently be possible to generate it from within the society itself. The villages in which subsequently participation evolved did show some signs of incipient internal leadership. On balance, however, there was a great degree of dependence on outside leadership and help. It is also found that variations in the degree of participation are partly attributable to the quality and kinds of leadership (such as political, governmental, social and so on).

3. Managing Common Assets

Another aspect that participatory institutions need to look at is the maintenance of community assets. As is well known, maintenance is an essential condition for transforming one-time income increases into sustainable ones. As the consequence of a long historical process of government intervention, population explosion and the collapse of institutions, rural communities have dissociated themselves from management of assets. They need to be retrained in it. This is an added organisational responsibility. This characteristic can emerge only by proper management of the surplus generated by CPRs. Expenditure on activities that ensure maintenance of higher level of output has to be incurred. For instance, tanks and water-conveyance systems have to be maintained and plantation has to be carried out on a continuous basis. Societal surplus could also be frittered away on mass consumption of the collective kind such as roads and religious places. It seems, therefore, that as a precondition for evolving participatory institutions, transfer of some resources to local non-market, non-governmental institutions and training of rural communities in their maintenance and management is essential.

Traditional wisdom ordains State management of public goods. Yet, it has been seen that in the CPRs where policing costs may be high, and where State management may involve illegal collections between government agents and local populations, it is beneficial both to the government and the local people if management is handed over to village - level societies. In actual practice, however, it is important to include adequate safeguards against misuse of such CPRs.

4. Awareness

Another important factor is awareness. The intent of participation depends largely on the degree and level of knowledge the people have about the programme. Most of the times, the absence of clear, correct and complete information, the local benefactors are not in a position to take any decision to participate or otherwise. The doubt creates fear and hence they prefer to remain passive spectators of various schemes.

The people assess the local situation and decide the course of action based upon their assessment of risks, uncertainties and likely benefits. The development planners, administrators and NGOs in rural development need to understand the cognitive map of rural people, the way they perceive the programme, its benefits, costs and risks. The person would decide to participate depending on how much it would contribute to his goal and then the larger interest of the community. He would compare the costs which have to be incurred and the relative gain which would accure to him and to the community before he decides to participate. Based on this assemption, Bryant and White proposed the following formula :

$$P = (B \times Pr) - C$$

Where,

P = participation which depends on

B = Benefits which are hopes to get,

Pr = Profitability that they will actually achieve the benefit, and

C = Costs involved to work for the projects.

This formula predicts that people will participate in an activity when they sense that their benefits outweigh the time and effort they are expected to spend. It should not be construed that this formula would make the trick of making of peoples' participation. However, it provides insight into formulating certain basic principles of participation which are :

- People organise best around the problems they consider most important in assessing needs and in planning initiatives to meet those needs.
- Local people make rational economic decisions in the context of their own environment and circumstances.
- Voluntary commitment of labour, time, material and money to a project is both an evidence of participation and a necessary condition for breaking patterns of development paternalism.
- Local control over the amount, quality and distribution of benefits increases the chances of the project becoming self-sustainable.

All the above mentioned areas give certain clues for identifying the following approaches which one might adopt possibly to enlist community participation;

- People should be made aware of the details of all the schemes.
- Identification of felt needs of the people / community need to be done with the involvement of the people.
- The planning process should be of "bottom-up" approach. The user of the programme should also be the decision maker in the programme. The present "top-down" approach should be changed to another approach called "bottom-up with support down", where the rural people plan their programme with the official support.
- Developing a sense of ownership and responsibility for a programme among the community members creates a feeling of possessiveness. To develop this, the outside agency should be prepared to gradually withdraw its role, delegating the responsibility and functions to the community.
- Subsidy should be given only to serve as supplement to the exsisting resources of the individual / community. A subsidy is most effective, if the individual / community takes it as a form of recognition of the efforts and that it is aimed at stimulating more effective development.

Need for People's Participation

Peoples participation at the local levels is necessary primarily to reduce the unequal distribution of power in the rural areas. People's participation at the local level would help bring about a redistribution of both control of resources and of power in favour of the rural poor. When the power structure is altered at the local level in favour of the excluded' and marginalised' sections of the people, they will get a new deal which would ensure a pattern of access of resources, services and the amenities in their favor. Apart from this, there are reasons why the people's participation is necessary at the local level. They are;

- To take note of the felt needs of the population;
- To mobilize local resources for plan implementation including peoples labour;
- To decrease the level of conflict during the planning and implementation stages;

- To increase the speed of implementation by securing the cooperation of the people;
- To increase the legitimacy of the authority; and
- To reduce popular resistance to decisions.

Politically, participation in developmental activities, besides being resource intensive paves the way for meaningful articulation of local demands. Planning thus becomes much more realistic and receives ready political support. From administrative point of view, local capability to govern local areas increases through sustained participation in local decision making.

As Julius Nyerere observed governments by themselves cannot achieve rural development. They can only facilitate it and make it possible. They can organize, help and guide. They must be able to control their own activities within the framework of their village communities. And they must be able to mount effective pressure nationally also. The people must participate not just in the physical labour involved in economic development but also in the planning of it and the determination of priorities.

A number of poverty alleviation programmes, viz. Integrated Rural Development Programme, Jawahar Rozgar Yojna, Drought Prone Areas Programme, Desert Development Programme have been launched by the Government of India. The evaluation studies of these programs shows failure in most of the areas. This has been mainly because the target groups have not been properly organised to assert themselves both in the formulation and implementation of these programmes.

It has also been observed that beneficiary oriented anti-poverty programmes have been implemented mostly through the official delivery mechanism. The active involvement of the beneficiaries is necessary to ensure that the assistance reaches the intended target group living below the poverty line and that the target group derives maximum benefits from the programmes. The ideal ssituation would be if the beneficiaries themselves take the initiative. But they have so far failed to assert themselves because of a number of factors. The G.V.K. Rao Committee constituted in 1985 to review the existing administrative arrangements for Rural Development and Poverty Alleviation Programmes suggested that the poor should be organised as members of formal institutions like cooperatives as well as members of informal groups. The motivated functionaries in the government organisations and banks can create awareness amongst the rural poor. Further, the Committee suggested that the objectives of any programme for the organisation of the poor should inter-alia be;

- To increase the level of awareness to the target groups in regard to the programme content and facilities provided therein under programmes such as IRDP, TRYSEM, DWCRA, etc.
- To encourage their participation in the planning and implementation of these programmes;
- To increase the bargaining power through group action;
- To promote cooperative and group action among the beneficiaries;

- To establish a feed back mechanism and appropriate forum where a constant dialogue could take place between government functionary groups at village level; and
- To make the beneficiary groups self-reliant, i.e., the group/individual increasingly learns to do for themselves/himself what was done for them previously by others.

Although the guidelines of the government of India call for an active involvement of the Panchayati Raj Institutions in anti-poverty programmes, infact, the involvement of these bodies is far too low. The study conducted by Rao revealed that the participation of PRI's was practically nil in IRDP because no specific role had been envisaged for these institutions in the IRDP except a cooperative role in the identification of beneficiaries through Gram Sabha's. Regarding the factors responsible for lack of peoples participation in IRDP, the study conducted by Sastry revealed:

- Beneficiaries have only a vague idea of IRDP;
- The IRDP has contributed to the generation/sustenance of additional avocations rather than additional incomes;
- Corruption among officials connected with its implementation;
- Non-viability of the assets/units sanctioned as also their poor quality;
- Shortness of the repayment period; and
- Cases of delays in grounding the schemes.

Community participation is gaining acceptance among policy-makers in India as an essential aspect of managing natural resources. This belated recognition is reflected in the government of India's somewhat half-hearted initiatives to facilitate community management of forests, and consider a larger role for users in the management of irrigation systems. Participatory approaches are recommended even for agricultural research, hitherto considered the exclusive domain of scientists. Fundamental changes, more drastic than in other sectors, are being attempted to increase community participation in watershed development. The common guidelines for implementing the projects supported by the Ministry of Rural Development, the major source of funds for watershed development in the country, require community participation. The district administrations, which receive the funds from the Ministry, are encouraged to use non-governmental organizations (NGOs) to implement the projects. The guidelines give the communities substantial freedom to decide how to use the money. Most significantly, the funds are placed under the control of watershed organizations. This is a major departure from the way the government usually works. Recently the Ministry of Agriculture agreed to join the Ministry of Rural Development in using common guidelines for watershed development.

Watershed development is attracting considerable attention because it is central to any strategy to develop rainfed agriculture or improve natural resource management. The development of rainfed agriculture is being emphasized as growth in its productivity has lagged behind that of irrigated agriculture. More than one half of the increase in agricultural production in the country has come from less than one-tenth of the districts, those with substantial access to irrigation[31]. Rainfed agriculture is considered to have untapped potential. The National Watershed Development Programme for rainfed agriculture was initiated largely to entend the benefits of the green revolution to rainfed areas. Soil and moisture conservation, the core element of watershed development, has always been considered to be fundamental in improving rainfed agriculture.

Participation in Watershed Development Programmes requires, at a minimum, that implementing organisations do not follow a blueprint for development, but involve the communities in analysing soil and water conservation problems and identifying strategies to alleviate them. Three aspects are critical to such a process. The ability of members to participate as community or to have a collective voice, decisions made jointly by the community and implementing organisations, and communities bearing a share of the costs. Participation requires community organisations. Whenever needed, implementing organisations need to work with communities to initiate collective processes. This process, which we refer to as social organisation, is critical in watershed development because it distributes benefits and costs unevenly and calls for shared sacrifice to manage common lands. If mechanisms are not put in place to ensure that watershed development is in everyone's best interest, those who bear the cost may undermine development efforts and projects will be wasted. Accordingly, as participation entails ownership by all participants, the agenda for watershed development needs to be developed jointly by communities and project implementers. A community sharing a part of the costs further strengthens its ownership of projects. These three key aspects of participation have implications for each other and overall level of community participation in the planning and development of watersheds and rural areas.

The government of India is under considerable pressure to improve the performance of water shed development programs. Successful programs have demonstrated that performance can be enhanced through community participation. The report of the Hanumantha Rao Committee, which was established to examine watershed development programme under the Ministry of Rural Development Drought-Prone area programme, attributed their failure to Adhoc planning and lack of community participation. Participation is the key ingredient of successful watershed development and the lack of it is one of the primary reasons for poor outcomes of watershed development in India, widely shared among policy-makers, donors, researches and other directly involved. The MORB issued the common guidelines requiring community participation on the basis of recommendation of the Hanumantha Rao Committee.

Constitution 73rd and 74th amendment Acts are considered as the mile stones in decentralized planning and governance. Well before enactment of these acts, through article - 40 of the directive principles of state policy our constitution had emphasized to take steps to organise the village panchayats and endow them with power and authority as may be necessary to see than as units of self-government. However empowering the PRIs does not mean that the rural poor would participate mechanically. In India problem lies with the development delivery mechanism as well as with the receiving system. We are always busy in improving delivery system rather we should have taken care to strengthen the receiving system so that the target community would become conscious to manage its own affairs after some years. People those who are disadvantaged in terms of wealth, education, ethnicity and gender are unable to identify the root cause of their poverty. And wherever they are able to do so, they do not have the required assets to intensify their activities. So it has become necessary to build capacity or engineer their ability and aptitude to overcome poverty.

Peoples' participation in planning and development for rapid rural development planning at the grassroots level is essential but the problem is who should plan and how it should be executed. Decentralized development planning in India emphasizes peoples' participation in the decision-making process. Planning is defined as "the thinking of a group of people poised for development and synchronizing their thoughts with their needs and abilities for anticipated action, which will result their welfare".

If we see our present decentralized process even after enactment of the Constitution 73rd amendment Act, the elected representatives are empowered to plan for development especially for the weaker sections. Even if it is meant for them, they do not have their options in choosing any specific programme, which they can manage very well. In this way the needy people are unable to take part in the local decision - making process because their representatives are doing it for them. The Gram Sabha is hardly organised to discuss about the proposed developmental activities. Though in the region where high literacy and awareness, and willingness to participate, to some extent has achieved its goal to involve people in the participatory development decision - making process, it would be biased to conclude that except a hand full of conscious people, others like those who are living under subsistence level would come and participate in the Panchayat, Block or District level seminars. By doing this again we are imposing the same elite and urban solution to the rural local problems.

Democracy is popular for peoples' representation, rather it should have been for their active and direct participation in local level development decision-making process. Representation and representatives kill the initiatives and innovative ideas with the people. Peoples' participation in planning or preparing "peoples' plan" would be possible only when they will discuss, identify the local needs and prioritize those identified needs to formulate micro-plans on the basis of a database which the people can follow. Participatory development is a process where the participants come forward to identify their needs, prioritize them, formulate small projects, implement the selected one, monitor, evaluate and also execute necessary follow up action. To involve people in the decentralized planning process, a reliable and authentic database is required which people can understand and analyze properly. political will and positive attitude for building awareness is also required for its successful implementation.

The most urgent need of the hour is to organise the rural poor and disadvantaged in terms of wealth, education, ethnicity and gender for social, economic, political and cultural development. Government agencies distinguish the rural poor as their target group and casually try to mobilize and muster them to deliver the goods without considering their behaviourial and attitudinal values. Mere mobilisation of people for certain programmes may encourage to build a dependent class. In the field of participatory planning and development NGOs are well ahead because they have started thinking in terms of with drawing from a locality after some years of intensive intervention which they think would bring in self-help and self-reliance to a community. In a participatory development process both the primary (reference people) and secondary stakeholders (the internal and external agencies like NGBO, government and donor) have equal responsibility to make a project successful.

People should plan for themselves in some key areas like education health, infrastructure and women experiment. According to development planner growth promotes equity, but it has failed in its mission so, this paradigm has been shifted to equity and social justice. Now it is the need of the hour to give the reference people a chance to participate directly in the development decision-making process. Peoples' participation should not be instrumental in providing information, presence in the meetings, to put attendance or even sharing the benefit. Rather it should be transformational participation such as participating in the discussion and decision-making process, formulate projects, implement, monitor, and evaluate the change experience for further planning and tap the potentiality of the rural poor and the local resources at their level and for their development. Collective decisions always score an edge over the individual or decisions made by only a selected few.

Bureaucracy and Participation

Administrators see a risk element in involving people in the development process. Participation may generate more expectations than can be ordinarily met. As Uphoff puts it:

"A commonly cited fear is that consulting with the intended beneficiaries about their problems and possible solutions will raise exaggerated expectations of what will be done for them. Rather than arousing such hopes, governments and technical agencies seem to prefer delivering benefits at a time and place of their own choice in order to keep control."

- Many bureaucrats are quite firm in their view that they alone know what development is and what is best for the people. According to this view, the people have no understanding of their problems and are simply incapable of improving their situation. Their disdain for knowledge and capacity of people is incurable. The only role for the people, which they are willing to concede is that of acting as recipients of benefits which development projects offer.
- Although there is no empirical basis to support the assumption that participation slows down the progress, yet among many administrators this indeed is a strongly held belief. Completion of the project on time, savings in expenditure and such efficiency considerations are then used as additional arguments to exclude beneficiaries from the planning and implementation processes.

- The administrators lack skills in working jointly with the people is the another reason for their reluctance. Pointing out this weakness in the manner of government functioning, a recent study concludes that : when it comes to working with local people to create self-sustaining development, evidence from the field indicates that large government agencies seldom have this capacity'.
- Too often, the local bureaucrats have been adverse to listening to ordinary citizens. Many of the projects created and managed by governments, moreover, leave little decision making to citizens and thus generate little popular participation and support. Frequently the best intentioned participative development strategies falter because they rely on a bureaucracy unable to respond to community needs and unwilling to rely on community skills and problem solving capacities.
- Not that administrators alone are to be blamed for a low level of peoples in the development processes.
- People themselves are generally apathetic and passive and because of their past unhappy experience with the government, their faith in the intentions of administrators seems to have diminished.
- Most people at the grass roots level are unaware of the possibilities of governments assistance that can be made available to them for developmental purposes.
- A virtual absence of community based organisations at grass roots level is the biggest stumbling block in the path of people centered development in the Third World countries.

However, there is some literature, which suggests that inefficiency is the cost of participation by the masses' in decision making. It is also to some extent true that workers control often lessens effectiveness and efficiency. Nonetheless, under some conditions efficiency and effectiveness are enhanced by high participation. In the rising view of the people's movements, participation is the breaking-up of the traditional relationship of submission and dependence, where the subject/object asymmetry is transformed into a truly open one of subject to subject in all aspect of life, from the economic and political to the domestic and scientific.

On the whole, there is a failure to ensure peoples participation. It has been recognized that no programme of the massive scale covering wide dimension of the country could succeed without peoples participation.

The development impulses are to emanate from the people and it is essential for the government to provide adequate environment and opportunity. However, this aspect has been missing to a large extent and a limited decentralization and involvement of people taking place in poverty alleviation programmes.This has brought to foccus the role of NGOs in bringing about people's participation in the rural development.People living under subsistence level need an altered approach to change their status and lead a decent life in society. Non-government development agencies are widely known for their approach towards rural development, particularly at the grassroots level, and in the inaccessible areas where they are very successful in initiating people's participatory process for positive change and development. They call their approach as holistic because they stay with the community, sit and discuss their problems to find out the most appropriate solutions. Like the government officers they hardly prefer to impose all expert solutions to the local problems, and always give due importance to local skill, resources that will play a positive and complementary role in all developmental activities. (Government agencies fail in their endeavor to involve people because they consider people as objects rather than subjects for change). The notion of considering development as creation rather then revolution is wrong and the colonial attitude of adopting problem - solution approach may not bring out a sustainable result.

NGOs and People's Participation

NGOs are in highly advantageous position to enlist the participation of people as they are at the disposal of the people, and as their duty hours are not time-bound. Nor are they governed by rigid set of rules and regulations as found in the bureaucratic administrative set-up and are flexible in their approach towards human development they adopt several ways and means to enlist the participation of the rural masses. NGOs are known for their commitment and dedication in the service of the poor. They place highest emphasis on the improvement of the quality of their work performance as nothing else would better establish their credibility than this. If only the government is willing to take into confidence the NGOs and is prepared to assign them a responsible role, if the NGOs and the government work together the rural transformation would become much easy. Bureaucrats might feel shy of learning from the NGOs. The true spirit and motivation for the cause of the poor should give way to such reservations. NGOs employ different strategies to enlist people's participation. Some of these strategies may be described as follows:

Inter-personal Communication: The very first and foremost thing which an NGO, does is to establish personal contact with the rural masses through inter-personal communication and contacts. For this the staff members of the NGO go to the people and discuss with them as they are and where they are. Anywhere people can be met and discussed without any reservation of time and place. This is just the reverse of what is happening in the government programmes. To win over the confidence and conviction and cooperation of the masses, initially the NGO may have to meet them several times, and more often.

Formation of Sangham (Village Association): After having established personal contacts with the people, the next thing the NGO does is to help them organise themselves into a group. This organisation is known by different names in different NGOs. But the aim is one and the same, namely, to involve the rual masses in all the aspects and process of rural development, especially in the sharing of the benefits in an equitable way. Again this is some thing which is unheard of in the government sponsored development programmes. In some NGOs this village-based body, sangham hold their own meetings once in a week. In such meetings, in the open discussion, the planning and decision making are done before undertaking a project. The same forum is used of for monitoring the ongoing projects and the evaluation of the completed ones. These meeting are also forums for setting the quarrels, disputes and other general problems.

Division of Villages into Clusters and Areas: If the NGO works in several villages, these villages are divided into clusters and areas. Each cluster consists of five to seven villages and each area consists of 20-25 villages. A cluster coordinator extends necessary direction and guidence to the sanghams in a cluster, an area coordinator coordinates all the activities of the villages in an area. Beisdes, there is an animator selected by the sanghams in each village. Members of a cluster of villages hold meetings once in a fortnight, and there is a quarterly gathering of the members of an area. These meetings and getherings are also forums for open discussion for planning, decision making, monitoring and evaluation.

Here the members learn from each other's experience. These are also forums for their self-articulations in public gatherings, a rare chance otherwise they are offered to. Normally such large gatherings are dominated by the elites. Here it is not the case. Some of the staff members of the NGO also participate in all these meetings. Reports of all these meetings are sent to the central office of the NGO. Continuous information thus gathered will serve as some sort of monitoring which serves as feedback for corrective measures to be taken about the ongoing projects and guidelines for future formulations and implementation of projects. NGOs also make arrangements for participatory evaluation of the projects during the implementation and after the completion.

Infrastructure for Interaction and Strengthening of Solidarity: Some NGOs have the necessary infrastructure facilities such as training centres for the sangham members to meet once a week. This weekly gathering, away from their own village environment helps in strengthening their solidarity. Their reactions and mutual interactions are other sources for further feedbacks for the NGO.

Regular weekly meetings with the field staff and the executives, visits of the executives to the villages and discussions with the people are regular features in NGOs. Contact with the rural people through mass meetings of large gathering on special occasions like international literacy, international women's day, environment day, May day are also being held by some NGOs.

However, in a large country of India's size, the rural masses cannot expect the direct participation as described above. But definitely they can expect it from the NGOs. If the NGOs and the government join their hands it is all the more easy. Any rural development project should lead the masses to self-reliance. Meaningful participation in rural development may be said to be on of the rights of the poor. On the whole it is said that rights are not granted but taken. Educating the rural masses on their lawful rights and helping them in taking the lawfully granted rights could be part of the awareness programme of the NGOs.

The aspect of peoples participation studied and observed by the researcher during the visits made to various agencies like manavlok,Grasp, Janarth,Abhinav Vikas Sanstha, Dilasa and Nirman are stated below and the role of MSSM in peoples participation is highlighted.

Participation in MSSM

MSSM worked for watershed development mainly in two talukas, viz., Aurangabad (villages Jadgaon, Hivra, Ladgaon, Mangrul, Tongaon, Satana and Karmad) and Jalna (villages Asarkheda, Kadwanchi and Buthkheda).

Awareness Building : In the initial stage of the project, involvement of the people depends on their understanding of the project and their assessment of the expected benefits. Three main aspects of collective learning were focused in this stage, viz., understanding the situation, analysing the problem, and identifying the options. These were done through formal and informal discussions and training events, as described below. In a typical project cycle, these steps broadly correspond to the pre-planning preparation stage.

Understanding the Situation : MSSM attempted to build a common understanding among the men and women in the villages on the situation of natural resources in their village through a dialogical process, which could be typically described as follows. Three to four meetings on the drought situation - a perpetual phenomenon - were held with the people. It was followed by screening of films and slide shows on natural resources degradation before men and women of the villages. A few days after the screening, a meeting was held with as many men and women (adults in the village) to discuss the message from the films and to decide the next step, which was named as "collectively understanding our village".

The villagers developed the understanding of village situation collectively using some tools of participatory rural appraisal (PRA). The fieldwork culminated in sharing of the maps and charts with almost all villagers at a village meeting. This was the basis for taking the next step of collective problem analysis.

Analysing the Problem : This was done in a series of informal meetings with large number of villagers. It was followed by a one-day workshop to consolidate the problem analysis. During the workshop, the villagers tried to identify their most pressing problems related to livelihood and their causes or reasons. These causes were grouped into two, viz., those linked to land and water, and those linked to individual or collective behaviour of the villagers. The exercise was mostly done using a base map - either a revenue map or the one prepared during the PRA exercises. Informal conversion on possible solutions was encouraged, but no ready-made solutions were offered. It invariably raised the level of curiosity among the villagers, setting the right condition to attempt the next step of exposure visit.

Identifying Options : Once the villagers became familiar with the problems of their village and possible reasons thereof, they attempted to identify options. Some villagers often have knowledge of related work done in other villages or in neighbouring areas. At this stage, exposure visit was planned for the men and women to see the possible ways in which their problems could be addressed. MSSM organised the exposure visits to areas where watershed work has been done and community organisation is functional and effective. During the field visit, interaction with the local villages was encouraged. Immediately thereafter, the experience and observations were shared in a village meeting and a broad development plan as to what could be adopted was prepared.

While the awareness building efforts aimed at a large proportion of the village population in the pre-planning stage, not all of them could take active part in the planning and implementation of the project. It was, therefore, necessary to identify small group or groups of people who could take more responsibilities than the rest. Such groups could be in form of committees set up with specific roles that they could fulfill responsibly, for example, watershed management committee, village development committee, sharamadan committee, water users' committee, forest protection committee, etc. Community organisation is about identifying responsible leaders, helping them identify their roles and capacitating them to fulfill the same. The approach adopted in the above projects had three major components, viz., identification of opinion leaders, involving them in operational aspects, and formalising institutional arrangements, as described below.

Identifying Opinion Leaders : The pre-planning state in all projects lasted for about 6 months to a year, which was sufficient to understand the strengths and potential of most opinion leaders in the villages. However, such identification was impressionistic, and therefore, required further testing, which was done by giving small tasks to these individuals and observing their performance. This provided considerable information on their leadership style and potential to shoulder various responsibilities.

Involving in Tasks : Most people were involved in the planning of the soil and water conservation works. The exercise, which lasted over a couple of months, involved visiting each farm to study its degradation status and to plan corrective measures of consultation with the owner of the farm. This task needed some technical skills of measurement of length, slope, depth, as well as using tape measure and calculator. Some youth and opinion leaders were trained in these skills. Similarly, technical skill training was done in tree-plantation methods, well-water measurement, assessing flood quantities, etc.

It was noted that the opinion leaders did not require much training in process or behavioural skills like communication, dialogue, conflict resolution, etc. Depending upon the interest and aptitude, the men and women were involved in various activities and tasks like supervision and quality control, measurement and payment, handling material, mobilising community, conflict resolution, etc. At individual level, they were given tasks with progressively increasing complexity, for example, from measurement of pits, to recording the measurement to computing volume, to preparing muster, etc. At collective level, the responsibility was increased by asking the villagers to take some decisions, while the project staff gradually reducing their own role in operational decisions. Such involvement helped not only in achieving the physical outputs, but also in developing their capacities.

Formalising Institutional Arrangements : It was found that the above groups of men and women worked effectively in informal way. But, for continuity of systems and for further development, a formal institution is desirable. Registration is one type of formalisation, but in order to ensure the long-term effective functioning, some other measures were tried. These included formation of small groups or task force federating into the committee, involving other groups like mahila and self-help groups, linking with other institutions like panchayats, etc. Some measures that were found effective in most villages are explained below.

Institutionalising Participation

Participation or involvement of people in any development initiative may be easier to achieve in some events or stages, but it is relatively more difficult to sustain throughout the programme, unless institutionalised. Some practical methods of institutionalising participation, as developed during the resolution mechanism, and maintenance fund, were found effective in most villages.

Broad-Basing Responsibility : Different interdependent roles were assigned to different people to enhance teamwork and synergy. For example, the tasks of quality check (supervision), measurement, and payment were allotted to different committee members, who shared their observations in the committee meetings (on larger platform). This also helped in mutual checks. Here, the underlying principle is that if the management of affairs is rested in the hands of larger number of people, as compared to a few individuals, there is a mutual check due to increased transparency. It also helped in allowing more space for more men and women to participate in the work, and thereby, increase their ownership in the project.

Conflict Resolution Mechanism : Minor conflicts kept surfacing in course of the implementation, such as provision of space for stacking material, exact demarcation of boundary between two farms, provision of land for alternative road around a dam site, etc., which were resolved by the villagers themselves in an amicable manner. The conflict resolution mechanism, with minor variation across the villages, constituted three main and progressive stages. The first stage would comprise of an attempt by the site supervisor to resolve the conflict through dialogue between the two conflicting farmers or groups. If they failed to arrive at a consensus, the issue would be discussed with the committee. If it could not be resolved there, a few elderly and respected persons would be brought in to mediate. An important aspect of this mechanism was that no compensation or special benefits were bestowed upon the conflicting parties from the project funds.

Maintenance Fund : As an element of ensuring continuation of the activity beyond project period, especially to address the maintenance and repair needs, maintenance fund is created in the watershed projects, mostly out of the cost saved on account of Shramadan. In the above projects, maintenance fund was collected from the farmers on the basis of their land holding, besides their shramadan contribution. This helped create an increased sense of ownership. The fund (to the tune of a few lakhs already collected in each village) would be managed by the people (committee) and could also be used for any development purpose, especially for development of women and children.

Inferences

Participation or involvement of people in any development initiative may be easier to achieve in some events or stages, but it is difficult to sustain throughout the programme. This is sometimes because of the limited availability of villagers, who are willing and capable of taking up the project responsibilities. This could be overcome by systematic awareness, motivation, training and capacity building[48]. The preliminary inferences, the conclusiveness of which needs to be established over a variety of socio-economic and socio-political milieu, are as follows :

- Capacity building of local people is essential to increase their participation in any development programme. The capacity building measures should be used gradually based on the response and rate of progress of the people.
- Shramadan, especially when expressed in numerical terms or percentages, is, at best, a poor indicator of people's participation. It does not represent the quality of involvement of people.
- Institutionalising participation and involvement of the community is essential for sustainability.
- Participation depends on agreeing to and practicing some core values by all the concerned. It is, therefore, necessary that the attitude and behaviour of the project staff are in consonance with these values.

In essence, the framework and the aforementioned inferences are drawn from the experience in watershed development projects, but they are applicable, with appropriate modifications or adaptations, to most rural development projects.

Participation in Manavlok

The activities of Manavlok are need-based, planned and implemented only after discussion with the target group. No program is implemented without the approval of the community and their participation. To bring about the community participation effectively in every activity. Manavlok has two important local level organisation viz., Krushak Panchayats' which consists of farmers and landless labourers and Bhumikanyamandals, which consists the rural women. Krushak Panchayat (KP) aims at democratic working, building unity among villagers, to provide a forum for their concerns, ensure the wise direction of funds and provide a semblance of bottom-up management. The members of this KP are also requested to participate in the planning and budgetting of all programs. The members also select the beneficiaries of the programs according to their needs and manage all the village level activities with the cooperation of Manavlok.

The members of Bhumikanya mandals come together, discuss, voice and solve their problems and join the KPs in the rural development programs. There is also an organisation Tarun Mandal' to organise rural youths in various activities.

Participation in Janarth

Activities of Jannarth are decided upon through a continuing dialogue with the villagers. This process helps in making the people vocal, sensitive, responsive and willing to take responsibility for their own development. The men and women are encouraged to involve in the planning and implementation stages in all the programmes. In the government sponsored programmes for watershed development, people are involved in every stage of work through registered village committees. Villagers come forward to participate voluntary, and even pay partially for soil conservation work on their land. There is a Mahila wing activity which aims at building confidence in women through their participation in Mahila mandal activities, village governance through their elected representatives. New women are getting together in informal groups to discuss common problem. In the urban slum activities of Jannarth self-help groups are formed wherein people discuss and take action for their economic betterment and overall development.

Participation in Grasp

Grasp is a gassroots action for social participation. The vision of grasp is to empower communities to manage their resource in a professional manner. To bring about the people participation in professional manner Grasp has some village level organisations like Village Watershed Committees (VWCs), Mahila Mandals (MMs). The VWCs focusses on building local capacities to maintain the resources created and work on forward linkages. The MMs are initiated to increase direct participation of women in economic and developmental programme. Grasp focusses on increasing community participation in the management of environmental resources through planned activities and execution of projects by the community groups. Grasp team approached the community by initiating a need-based dialogue with the village community on specific issues, such as self-help groups formation, installing micro-irrigation schemes in groups, provision of improved agricultural packages, enhancing livelihoods through optimum utilisation of land and water.

During implementation of any kind of programme the community contributes in the form of labour (Shramdan). The project reimburses the contribution to the community in form of village fund, which the community uses for operation and maintenance of community works. This creates a sense of ownership of the programme.

Participation in Nirman

This organisation is build up on the support of and in association of the community. The organisation tends for awareness, planning and implementation of participatory self-initiatives while working with the rural and urban communities. Nirman believes in the capacities of the villagers. Village youths, called as village mobilisers are provided on job training in participatory planning, implementation, and monitoring. The village mobilisers carry the movement of participatory resource management to many other villages. These skills have helped them to earn their living and status in the society. The village institutions such as self-help groups, panchayat, user's groups, mahila mandals, yuvak mandals, etc. are promoted and strengthened by various efforts of Nirmaan.

Participation in Abhinav Vikas Sanstha (AVS)

One of the major approach of this organisation is community organisation, participatory planning and implementation. To facilitate this approach AVS has formed various groups at village level. Youths, farmers and women are promoted to form self-help groups. At present there are 30 SHGs in the village. The groups meet regularly share their feelings and have established their own identity. Krushi Mandal is formed with the farmers for information, sharing and transfer of knowledge for agriculture promotion. The farmers interact with the experts from various fields and are benefitted by modern technologies, critical inputs and hand holding support. AVS concentrated hard to organise youth. They are promoted to address their our problems. These youths have voluntarily constructed a bus stop costing Rs. 15,000 at Soegaon through Shramdan. With the help of youth self-help groups five members have established their own business.

Participation in Dilasa

Dilasa formed 310 groups of women to take up economic activities suited to their skill, aptitude and the local conditions. It also formed 22 women's co-operative dairies in the district. The organisation is also striving for bringing about children's participation in school sanitation programme by chaning their attitudes.

With the above description of the participation aspect in the NGO's working in Marathwada Region, it can be firmly stated that for the success of any project or activity either of the government agency or the voluntary agency involvement of the people, their desire, their need in all the stages of the project is very essential.

"The enormous task of social and economic revolution can be carried forward successfully only with the participation of the people. Obviously, the administrator has to go to the people. He has to identify and maintain a certain aloofness - his task would become easier if the administrator is genuinely dedicated to the cause".

References

Pathak H., Citizen and Public Administration in the Indian Journal of Public Administration, Vol XXI, No : 3 II, Special Number - Citizen and Administration II, New Delhi IIPA, July-Sept. 1975. P. 575.

Eldersveld, Jagannathan and Barnabas, The Citizen and the Administration in a Developing Democracy, Illinois Scot Foresman Co, 1968, p. 4.

Robin William, A. The Civil Service in Britain and France, London, The Hogarth Press, 1956.

Dr. P.M. Bora, Peoples Participation in Food Administration In India : A Study of an Indian State 1982 A Janta Publication. p. 201-202.

Robin William, Op.cit, p. 5-6.

The Concept of Active Citizenship was Emphasised by Shri V.T. Krishnamachari in the opening session of the Fifth Annual Conference of the Members of IIPA. See the Report Administration and the Citizen',] Proceedings of the Fifth Annual Conference of the Members of IIPA New Delhi, IIPA (1961), p. 4.

M.L. Santhanam : Community Participation for Sustainable Development, The Indian General of Public Administration, Special Issue on; Towards Sustainable Developmet of Society - Imperative and perspectives, July-Sept. 1993 Vol XXXIX, No : 3 (ed) p. 413.

Ibid. p. 416.

Cohen, C. Demoracy (New York : Free Press 1971.) p. 7.

S. Nagendra Ambedkar : People's Participation in Rural Development. pp. 98-99.

Kanchan Chopra : Gopal K. Kadekodi, M.N. Murthy, Participatory Development : An Inroduction-A Fresh Look at Rural Development Strategies. pp. 17-18.

S. Paul, Community Participation in Development Projects, The World Bank Experience' in Readings in Community Participation Washington D.C. The World Bank, 1987.

Piciotto, Robert] 1992 : Participatory Development - Myths and Dilemmas, WPS 930, The World Bank, Washington.

H.J. Eysenck, et al., Encyclopaedia of Psychology, Vol 2, London, Search Press, 1972.

J.M. Cohen and N.T. Uphoff, Participation's Place in Rural Development : Seeking Clarity Through Specificity, World Development, Vol 8, No. 30, 1980, pp. 213-216.

Uma Lele, The Design for Rural Development : Lessons for Africa, Baltimore, John Hepkins, 1975.x

ACC Task Force on Rural Development, Report of the Third Meeting of the Working Group on Programme Harmonization, Rome, 1978.

Baumgartner, Ruedi (1989); Participative and Integrated Development of Watersheds : Reflections on Experiences from an Indo-Swiss Collaboration, Swiss Development Corporation, Banglore.

John M. Cohen and Norman T. Uphoff : Rural Development Participation Concepts and Measures for Project Design, Implementation and Evaluation, Ithica (Cornell University, 1977) p. 7.

Kanchan Chopra, Gopal K. Kadekodi, M.N. Murthy : Participatory Development, Emergence of Participatory Institutions : Conclusions and Policy, Chap. 7, p. 139.

Kanchan Chopra, Gopal K. Kadekodi, M.N. Murthy : A Fresh Look at Rural Development and Strategies, Participatory Development : An Inroduction : Sage Publications - New Delhi. pp. 139-141.

The Optimal Sharing Arrangement Between the Government and Society can be obtained as a Solution to a Two-Person Co-operative Nonzero Game. If the resources are manage by the Government and People Differently. The Production functions for the Government and Society Can be Specified as Q_1=F_1] L_1 R_1 Q_2=+] R_2 F_2] L_2 R_2 where L, = Labour input Q_1 = Produce i= 1 Government i = 2 Society + R_2 = Index of People's Participation.

D.D. Gow and J. Vansant, Beyond the Rhetoric of Rural Development Participation : How can it be done ? World Development, Vol II, 1983, pp. 51-54.

Planning Commission, Report of The Working Group on District Planning : New Delhi, Planning Commission 1984. p. 83.

Mohit Bhattacharya, Bureaucracy and Development Administration, New Delhi, Uppal Publishing House, 1991. p. 88.

Juluis K. Nyerere, on Rural Development Impact, Vol 15, No. 3, March, 1980. p. 88.

S.N. Ambedkar, In Rural Development Programme : Implementation Process (Jaipur, Rawat, 1994).

Government of India, Report of the Committee to Review the Existing Administrative Arrangements for Rural Development and Poverty Alleviation Programmes (New Delhi : Department of Rural Development, 1985) p. 55.

Y.N. Rao, Administration of Rural Development Programme' : A Study in Ranga Reddy District of Andhra Pradesh, Journal of Rural Development, Vol 9, No. 6, p. 990.

K.R. Sastri, Factor Affecting Rural Development Participation with Special Reference to IRDP and NREP, Journal of Rural Development, Vol 9, No. 6, 1990. P\p. 1078.

Government of India] GOI 1992 : National Watershed Development Proejct for Rainfed Area] NWDPRA : Guildelines, Ministry of Agriculture, New Delhi. p. 1091.

Fan, Shenggen, and Peter Hazell] 1999 : Are Returns to Public Investment Lower in Less Favoured Rural Areas ? An Empirical Analysis of India, EPTD Discussion Paper No. 43, IFPRI, Washington.

Sinha M.S.] 1995 : Review and Status of Rainfed Farming and Watershed Management in India in P.N. Sharma and M.P. Wagley eds). The Status of Watershed Management in Asia, UNDP/FAO Katmandu, Nepal.

Seth, S.L. (1996) : The National Watershed Development Programme For Rainfed Areas (NWDPRA) : Restrospect and prospects' in Jensen, TR, et al] eds Watershed Development - Emerging Issues and Framework for Action Plan for strengthening a learning process at all leve, proceedings of DANIDA'S International Workshop on Watershed Development, Hubli and Bangalore, India, December 1995, WD CU, New Delhi

Hinchucuffe et al] eds] nd : Fertile Ground - The Impacts of Participatory Watershed Management, Intermediate Technology Publications, London.

GOI] 1994a : Deshpande R.S. and N. Rajshekhar (1995) : Impact of National Watershed Development Programme for Rainfed Areas NWDPRA Agro Economic Research Centre, Gokhale Institute of Politics and Economics, Pune Maharashtra.

Sitakanta Sethi, Participatory Planning for Rural Development, Ashwattha-Quarterly Journal of YASHADA, Vol 3, No. 1-3, Jan.-Sept. 2000. pp. 16-18.

Norman Uphoff, Fitting Projects to people' in Cernea (ed) Putting People First : Sociological Variables in Rural Development, New York : Oxford University Presss, 1985. p. 381.

Robert Chambers, Rural Development : Putting the Last First (London : Longman, 1983).

Benajamin U : Bagaden and Francis, F. Korten, Developing Irrigators Organisation : A Learning Process Approach' in Cernea (ed) Putting People First. p. 85.

Ramesh Arora : Summing Up' in Rakesh K. Arora] (ed) Politics and Administration in Changing Societies : Essays in Honour of Professor Fred W. Riggs, Op at, p. 350.

C. Francis, Rural Development, People's Participation and the Role of NGO's. Journal of Rural Development, Vol 12] 2 1993, NIRD Hyderabad (India). p 205.

Ibid, p. 210.

Although the experiments cited here are from Marathwada the operational area of MSSM, similar efforts were made in other parts by other voluntary organisations, esp under programmes by AFARM, AFPRO WOTR and IGWDP.

J. M. Gandhi and Yugandhar Manadvkar : Involving Communism in Watershed Management Challenges Approaches : Paper Submitted to the International Seminar on "An Integrated Approach for Strengthening and Protecting Water Sources" at Pune on Sept. 27-28, 2001.

MSSM is presently experimenting with the idea of promoting watershed development through Gram Panchayats capacity building of Gram Panchayats members and functionaries from three selected villages has been initiated in Jalna in Collaboration with the Zilla Parishad.

This Methodology originally evolved by MSSM for Pragmatic Operational Planning, has been refined further. It is now being used by Watershed Organisation Trust] WOTR and NABARD as the official and mandatory planning tool called Net Planning' in the IGWDP.

Watershed Organisation Trust] WOTR Ahmednagar has developed a 24 step capacity building methodology for use in the Indo-German Watershed Programme (IGWDP). It has proved quite useful in the past in increasing the technical skills and knowledge of the rural people.

CHAPTER-6

Voluntary Agencies and Empowerment of Rural Women

Introduction

This chapter studies the definition of empowerment, rationale and process of women's empowerment. The issues in women empowerment are enlisted, giving the various attempts made by the government towards it. Lastly, the role of NGOs in the empowerment of women in Marathwada is described in this chapter.

During the past half-a-century, India has witnessed a positive transformation in women's empowerment and economic development while retaining a great diversity in its political and social system. Although India still has to go a long way in attaining gender equality and gender justice, no one can deny that its efforts towards redressing gender inequality are more pronounced than in many other countries. We have had about five decades of planned economic development. It is ironical that we are still trying to tackle basic needs for a minimum standard of living such as drinking water, shelter, sanitation and employment. In all these and many other essential aspects of economic development, women have been left far behind.

The present socio-economic and political changes of the developing world, rigid customs and traditions are giving way to a new order. This period of transition offers an unprecedented opportunity for government and voluntary agencies, both national and international, to help women prepare for their new roles. To go through this period of transition successfully socio-economic variables will have to be taken carefully into account. The most important of these is to ensure the economic independence of women as a right of individual citizens. The factors which will promote this are increased earning opportunities ensured by better health standards and appropriate educational-cum-training facilities[2]. However, basic to all this is the needed change in attitude - both individual and social - for promoting a new pattern of life. This requires a changed approach of individuals within the family and also a new approach to the study of family problems and services needed to enable the family to be the main instrument for social change. The fact which must not be underestimated is that much opposition will have to be overcome in all the areas mentioned above, importantly women are able to participate actively and fruitfully in the life of their communities.

Integrated, multipronged development approaches, permit the development of closer-to-real-life, problem-centred analytical models. Vision to collaborate in developing co-operatives strategies for addressing development problems is needed. Gender justice should also include consideration of differential access to power, resources and opportunities, as well as autonomy of action and self and social valuation.

Empowerment of women is a critical issue that is being discussed all over the world. This concept has roots in the women's movement throughout the world and particularly by third world feminists. It is since the mid 1980's that this term became popular in the field of development, especially with reference to women. In India, it is the Sixth Five Year Plan (1980-810 which can be taken as a landmark for the cause of women. It is here that the concept of women and development was introduced for the first time. India's rapid development is intrinsically linked to the social, economic and political empowerment of every citizen of this country. Ensuring gender equality by removing gender discrimination is the key to real empowerment.

In rural development women play a significant role, but empowerment of rural women has not taken place. In this process of empowerment of rural women NGOs can play a very significant role.

The Year 2001 was declared as the Year of Women's Empowerment for enhancing the status and sustainable achievement in women's empowerment. Therefor, the Prime Minister had appealed to all the citizens to participate actively for promotion of women's access to quality education, credit, health care, employment and social security. In this context, the Government of India is developing different programmes and strategies, establishing different organisations and creating various legal provisions. The UN Secretary-General (2000) announced a new United Nations initiative to demonstrably narrow the gender gap in primary and secondary education by 2005. He said that educating girls is a social development policy that works and has immediate benefits for nutrition, health, savings, and reinvestments at the family, community and ultimately at the country level.

Hon. Justice J. S. Verma, then Chairperson of NHRC stated on the occasion of the international women's day that there were serious violations of women's rights. They don't need just equality of opportunity, but more, in order to achieve equality through gender justice. If gender issues are all addressed, the three un-freedoms' namely illiteracy, malnutrition and lack of health care can be overcome as quoted by the Nobel laureate, Dr. Amertya Sen.

The term empowerment has the most conspicuous feature containing the word Power', which means control over material assets, intellectual resources and ideology. The process of challenging existing power relations, and of gaining greater control over the sources of power may be termed as empowerment quoted empowerment as an active, multidimensional process, which enables women to realise their full identity and powers in all spheres of life. Power is not a commodity to be translocated, nor can it be given as alms. Power has to be acquired and once acquired, it needs to be exercised, sustained and preserved. Bhasin explained the women's empowerment comprehensively and elaborated its meaning in different dimensions as shown in the following figure 2.

Fig. 2: Conceptual Frame-work of Women's Empowerment.

Promoting qualities of nurturing, caring gentleness

Understanding of the importance of human values, rights and privileges

Enhancing women's self-respect and self dignity

Creating and Strengthening women's groups and organisations

Conceptual frame-work of women's empowerment

Making women economically independent and self reliant

Reducing women's burden of work, specially within the home

Greater ability to overcome restrictions imposed by customs, beliefs and practices

Helping women fight their own fears and feelings of inadequacy and inferiority

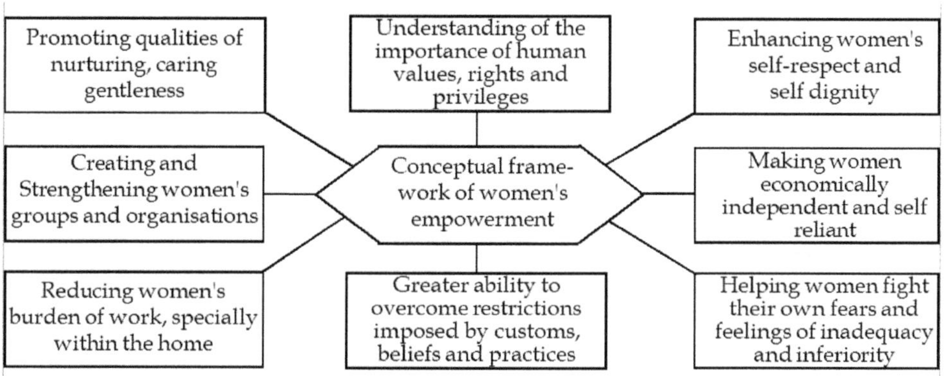

Source : K.C. Vashistha and Sashi Malik : Some strategic efforts towards the empowerment of women, University News, Vol 5, Feb. 4-10, 2002, P. 12.

Moreover, it can be said that empowerment is a process of awareness and capacity building leading to greater participation, to greater decision-making and control, and to transformation action.

Rationale for Women's Empowerment

Quoting from a UNFPA report, The State of World Population 1992', the Newsletter of Bernad Van Leer Foundation says that there can be no sustainable development without development for women, because it is women who contribute most for the development of children. The women have and will lead us in our search for a world free of violence and war. Women have led the peace and ecology movements in many parts of the world. Historically and even today women take care of the basic needs of society like food, fodder and fuel, they have creative to move in tune with their own nature. They have also been creative and nurturing.

The world Conference convened by the United Nations Decade for Women aptly remarked : "While women account for half of the world's population and perform two-thirds of the hours worked (though are recorded as working only one third of these hours) they receive one-tenth of the world's property registered in their name. These inequities are either ignored in development planning and policy formulation, or reinforced through specific development projects and policies."

The National Perspective Plan for Women 1988-2000 (1988) which has made a free, frank and objective analysis of the impact of developmental plans and programmes of Indian women, highlights the pathetic profile of women in India: There is continued inequality and vulnerability of women in all sectors-economic, social, political, education, health care, nutrition and legal. As women are oppressed in all spheres of life, they need to be empowered in all walks of life.

Further the Indian Constitution has guaranteed, Equal rights and equal opportunities to all its citizens, irrespective of sex, age, race or religion'. The rights are guaranteed by law, but yet to be legalized in custom.

Improvement of the status of women and their access to family planning services, make a triple contribution to sustainable development such as they make their own contribution to the quality of life and eradication of absolute poverty, they contribute to economic growth, by raising the quality and skills of the workforce and slowing down population growth thus reducing the burden on the environment which will improve sustainability. In brief, it can be said that women are to be empowered in the search for a safe environment, economic and social justice, adequate reallocation of resources, the survival of all species and the common goal of healthy planet in which the future generation can flourish.

Process of Women's Empowerment

Empowerment is a process both individualistic and collectivistic since it is through involvement in groups that people most often begin to develop their awareness and the ability to organise to take action and bring about change. Empowerment through awareness building, capacity building and organising women, leads to transformation of unequal relationships, increased decision-making power in the home and community, and greater participation in politics. Women's development can be viewed in terms of five hierarchical levels of equality i.e. Welfare, Access, Conscientisation, Participation and Control. Through these five stairs women may reach the status of empowered women. (Fig. 3).

It is convincingly noted that it is essential to design, implement and monitor, with the full participation of women, effective, efficient and mutually re-informing gender sensitive policies and programmes, including development policies and programmes at all levels that will foster the empowerment and advancement of women.

India has witnessed a positive transformation in women's empowerment and economic development in the past half a century. The transformation in women's empowerment comes even while retaining a great diversity in its political and social systems. Although India still has to go a long way in attaining gender equality and gender justice no one can deny its efforts towards redressing gender inequality is more pronounced than many other developing countries.

Fig. 3: Process of Women's Empowerment.

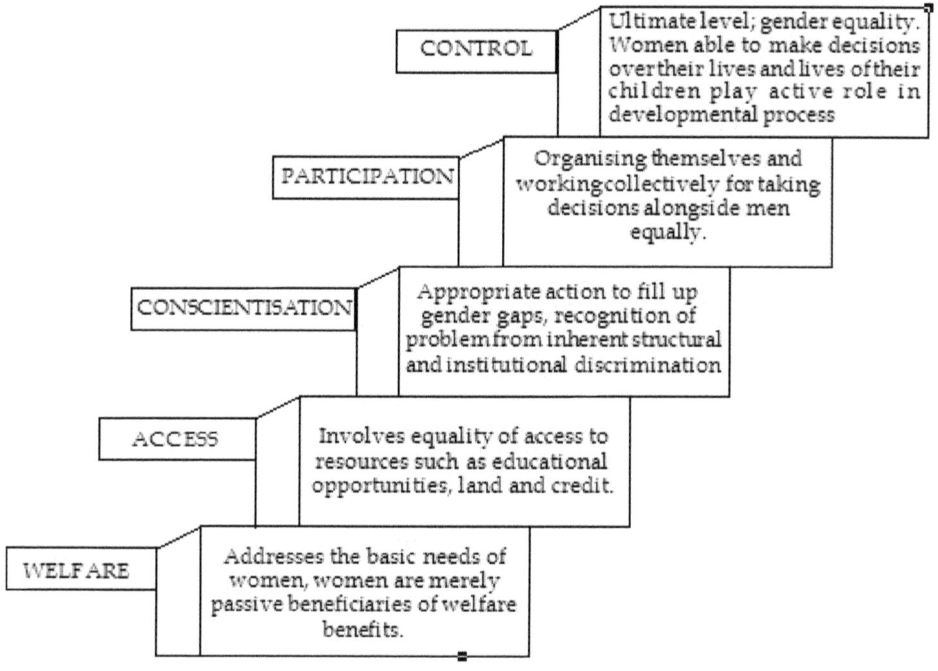

Source : K.C. Vashistha and Sashi Malik : Some strategic efforts towards the empowerment of women, University News, Vol 5, Feb. 4-10, 2002, P. 12.

State initiatives have been a part of our five year plans in the form of various projects for women, for rural development and for scheduled castes and scheduled tribes in the form of a special component plan and Tribal Sub-plan. Unfortuntely government tends to measure "progress" only in the terms of money outlays or infrastructure without taking into account whether the ground parameters are actually changing.

Issues of Women Empowerment

Money is not the main problem in the context of women empowerment but apathy, fear, ignorance and vulnerability as women within patriarchal patterns of social behaviour is a weeks. Integrated multipronged development approaches, likely will take on gender justice consideration of differential access to power, resources and opportunities as well as autonomy of action and self and social valuation.

The report of the National Commission on self-employed women 1988, provides ample evidence of the contribution of the government mainstream policies to the swelling of the unorganized sector, in which women workers have virtually no protection and are in no position to resist vested interests. After eight five year plans over the last five decades even the government has now come to concede the conventional strategies for women's betterment.

It has been conceded by the government NGO's and third party observers, that although income levels have risen and progress (in terms of more number of radios, bicycles, watches, savings account in banks, industrial activity, longevity, rise in literacy ratio) had indeed taken place, these have not meant that women's status and empowerment have also improved. They may have jobs, better schooling but social perceptions of women as second-class citizens have not changed propotionally. Women too think of themselves as lesser beings crimes against women (domestic violence, and battering, rapes and seductions, molestation, dowry deaths) have seen a relentless rise. In other words economic prosperity has not translated into real improvements in the status of women, in qualitative terms.

Empowerment in the sense of bringing women into the mainstream of national activity as equal partners alongside men has not materialized even after 50 years of planned progress. The reality that women face is not fully reflected in the economic models of development because growth in material terms can take place without diminishing inequalities and inequities.'There is not a single country in the world-not one - where men and women enjoy completely equal opportunities therefore, we must change attitudes and policies'.

Women constitute 70 per cent of the world's poor population of 1.3 billion. They produce 50 per cent of food worldwide but receive only 10 per cent of the income. In terms of every set of indices of development and socio-economic status, women have fared worse than men, in all region, in all strata of population; in work, employment, earnings, education, health status and decision making powers there is a clear difference between male and female entitlements.

Feminization of Poverty (a rise in proportion of women among the poorer strata) : It has been noticed as a phenomenon in most developing countries. The men move up faster during development, grabbing greater opportunities that women are unable to access because of gender handicaps, for instance manual work like hand pounding of rice, fire wood cutting and sale etc., have been replaced by mechanization, depriving the women of their source of income. The development leaves the women reluctant. This makes the percentage women of extremely poor category disproportionately large, necessitating special attention for them.

Earnings : Women's average earnings are consistently lower than those of men (Table 19). In the unorganized sector which accounts for 94 per cent of economically active women, earnings are ever lower.

Table 19 : Average Wage Earnings Per Day (in Rs.)

Age	*Rural*	*Urban*	*Male*	*Female*
-	Male	Female	-	-
15-59	16.09	10.85	23.72	17.36
All ages	15.04	10.11	23.30	16.86
Casual wage labour	-	-	-	-
(age 15-59)	10.53	5.11	11.89	5.30
All ages	10.27	4.89	11.09	5.29

Source : National Sample Survey Organisation (NSSO).

Health : Several studies have recorded a gender asymmetry in the utilisation of health services. During illness, fewer women seek and receive treatment as compared to men, for every 3 men who avail of hospital facilities, there only one woman who does so. Data from PHC's indicate that in some areas the skewness is as high as five men for every one woman.

Health is a function not only of medical care, but also of economic and cultural factors - Women neither have the time, mobility, child-care assistance, nor leisure to travel to health centres for medicines or treatment. In addition, an ideology that glorifies self-effacement and suffering for women makes them more inclined to put up with pain and ill-health, rather than demand attention and rest.

In terms of food intake, women suffer greater deprivation intra familially (women eat last as part of culture, self-effacement). If there is insufficient food men and children feed first and women do without. Because of gender bais in the allocation of food, there is a higher nutritional caloric deficiency in women, girls than in men/boys. The UN development fund for women (UNIFEM) estimates that 50 per cent of the women in Asia and Africa are malnourished.

Discrimination in entitlement begins with birth - Mortality rates are higher for girls than for boys in the 0-4 age groups, and 1.4 million girls are said to be missing in the age group 0-6 due to various manifestations of unequal treatment and gender bias.

Education : Out of the 960 million illiterate persons in the world, 640 million or two thirds are women. In every age group, literacy figures are lower for females compared to men. In India out of 428 million illiterates, women account for 275 million, with male and female literacy rates being 63.8 and 39.4 per cent respectively.

Policies : Political space has been monopolized by men. Representations of women among members of parliament and State legislatures have never exceeded. Women's representation at the higher rungs of decision-making position has also been consistently low.

Socio-Cultural Factors : The cultural construct of son-preference adds to the psychological diffidence of the female child. Dowry deaths are related abomination According to India-National Crime have been inn last three years 24,771 dowry deaths records Bureau of the Home Ministry.

During 1995, United Nation Development Programme (UNDP) developed gender-related indices for assisting human development, in the form of the Gender Development Index] GDI to take into account disparities between men and women, and the Gender Empowerment Measure] GEM to include the representation of women in political decision making. Globally there is not a single country that has a GDI higher than the Human Development Index. (Table No. 20)

Human Development Index and Gender Development Index in the Major States of India : Table 20

	Life Expectancy (age 7+, 1991)	Literacy 91-92	HDI 91-92	GDI
Kerala	72	90	.603	.565
Punjab	66.4	59	.529	.424
Maharashtra	64.2	65	.523	.492
Haryana	62.9	56	.489	.370
Gujrat	60.1	61	.467	.437
West Benga	161.5	58	.459	.399
Himachal Prades	63.6	64	.454	.432
Karnataka	61.9	56	.448	.417
Tamil Nadu	62.4	63	.438	.402
Andhra Pradesh	60.6	44	.400	.371
Assam	54.9	53	.379	.347
Orissa	55.5	49	.373	.329
Rajasthan	58	39	.356	.309
Bihar	58.5	38	.354	.306
Madhya Pradesh	54	44	.349	.312
Uttar Pradesh	55.9	42	.348	.293
INDIA	59.4	52	.423	.388

Source : India : The Road to Human Development, UNDP, P.47. States are ranked in descending order of HDI. A gender development index of 1.0 represents equality between men and women; figures lower than 1.0 indicate the extent of disparity as per measures devised by the UNDP Human Development Report, 1995. Delhi, 1997.

Violence against women has its roots in subordination of women at the social level and their vulnerability has not decreased but increased as consequence of social disruptions backlash hostility to women self assertiveness and growing awareness, and a general degeneration of values. Every seventh minute, the National Crime Records Bureau records a woman subject to some criminal offence or the other.

Issues of the Rural Women : In spite of gender equality being one of the constitutional mandate women lay behind men in terms of every set of socio-economic indicator. Inequality in the wages system shows that wages are lower for women, the rates are also lower for rural women. Table No. 19 shows the difference between urban and rural women's wages.

The comparison of work hours of women and men across the globe show that women work longer hours than men in both the market and non market activities and this is especially true in rural areas.

Literacy rates for rural women likewise are lower than for the total female population, 30.9 per cent and 32.9 per cent respectively. What is significant is that the rural-urban gap in female illiteracy has increased by 3 per cent during the last decade. The figures were 69.1 per cent for rural female and 36.1 per cent for urban female respectively.

The Mortality rates both Infant and maternal are significantly higher and for the rural population. It is also seen that there is an increase in infant mortality rate in the dalits than in other Hindus. Only 5 per cent of rural population is covered by proper sanitation facilities, while in urban population 50 percent is covered. Life expectancy rate shows a difference it is 58 years in rural population against 64.9 years for urban people, (1989-93).

SC/ST Women in rural population : One Quarter of India's population belong to communities known as scheduled castes and scheduled tribes (SC/ST). Together these two groups account for 206 million persons - a population larger than that of several countries of Europe put together. Within this segment the rural population accounts for 112.39 million (SC) and 62.75 million (ST) or 81 and 92 per cent respectively. These sections are among the poorest of the poor in the country even after half a century of planned development and inputs of several constitutional provisions specifically spelt out for their betterment.

Rural women from scheduled castes and scheduled tribes carry a triple handicap - one as women, second as rural person and thirdly as scheduled castes or tribals. It is for these women, there should be a focus on an alternative strategy for development of these women.In rural areas 44.7 per cent of the SC's fall below the poverty line among the ST's- 52.6 per cent fall below the poverty line.

In spite of article 46, of the Directive Principles of the Indian Constitution, which specifically spells out the States commitment to promote the special care education and economic interests of the weaker sections of the people in particular of these of the SC/ST. NGO's engaged in women's development have pointed out that "a most distressing dimension of educational disparity is presented by more than 90 per cent illiteracy among the rural female belonging to scheduled castes and schedules tribes".

State Initiatives Towards Women Development

For the first 25 years of development planning in India it was assumed that the general benefits ultimately be shared by all sections and that whatever benefits accrued to the men (or to the family as a reckoning unit) would also percolate to the women and result in better status and gender equality. This however turned out to be a faulty assumption, as periodic assessment at the end of each plan period showed.

In 1971, the Government appointed a commission on the status of women in India. Its report, Towards Equality' led to debates that highlighted the need to view women not just as targets of welfare policies, but also as "critical actors of development". However the planners approach remained largely unchanged, and by the end of the Fourth Five-Year Plan (1969-74) it was clear that the growth effect was not trickling down to the weaker sections of the population, including women, scheduled castes and scheduled tribes. A Minimum Needs Programme was therefore incorporated into the Fifth Plan design. Since the household continued to be targeted as the beneficiary, the approach itself was based on the concept of patriarchy. There was no mention of women's specific needs in the areas listed for priority attention under the programme.

The observance of an International Women's Year in 1975 resulted in attention being focussed on women, their problems and needs and in April 1975, the Lok Sabha passed a resolution to initiate comprehensive programme for women. A separate Ministry for Women and Children was created in 1984. Issues pertaining to rural women's advancement were identified by three planning groups of the Government of India. The National Committee focused on the role and participation of rural women in agriculture and rural development.

The Sixth plan document began, again with an admission that the weaker sections (SC/ST, women and the rural poor) were still lagging behind in terms of incremental development. A chapter on women and development was included for the first time in the Sixth Plan document.

Targeting rural poor women in particular, the plan declared that, a fair store of employment opportunities would have to be created through poverty alleviation programmes". The assumption continued to be that employment and income generation would automatically lead to improvements in the status of women.

Various Rural Development Schemes

The various schemes chalked out for women development in the Sixth Plan period with emphasis on the economic development were IRDP, DWCRA, TRYSEM, 41 items under employment generation schemes which include National Rural Employment Programme (NREP), The Rural Landless Employment Guarantee Programme (RLEGP), the Jawahr Rozgar Yojna (JRY) the Shram Shakti Development Guarantee Scheme and Integrated Area Development Programme (IADP).

IRDP : The IRDP was to cover 20 million families using package of subsidies and institutional credit, out of which 30 per cent was to go to women. A mid term review revealed that only 12 per cent of the target could be reached. The IRDP can be described as an example of tangled thinking. Subsidized loans were given to acquire assets, for instance buying a buffalo, which could not be sold piecemeal when emergency occurs or the animal dies of disease, there would be no permanent self sustaining improvement (in the absence of supportive veterinary services or training). Tailoring training was made available, the sewing machines were donated, but it was discovered at the end of the training, there was no market for a trained woman's sewing skills in the area (villages were too poor to generate business). She had to pawn or sell the machine to survive. The basic vulnerability of women, which was the cause of poverty was not tackled or even addressed. Training therefore, often turned out to be a non-viable economic activity.

Under the NREP and RLEGP projects for creating community assets, there was a mandate to give preference to SC/ST's, but in subsequent assessments made by the planners themselves, it was concluded that none of the projects had delivered the expected results in the form of overall betterment of women status in these communities.

The reasons were as follows :

- Lack of coperation among the officials entrusted with implementation.
- Procedural hassles that most of the illiterate applicants found intimidating.
- Most important, failure to involve the beneficiary population, as participants and treating them as subjects / targets with decision making in the hands of the officials.

Inspite of the actual outlay of Rs,1,00,800 million by the Minority of Rural Development under the seventh plan, no appreciable improvements in the status of women followed.

DWCRA was introduced in the second half of the 6th plan, especially to target rural women. Beginning with 50 districts, this revolving giant scheme had by the end of the Eighth plan, covered 291 districts with the central and state governments, and UNICEF each contributing one third of the seed money. The women beneficiaries were encouraged to form thrift groups, which were expected to work better than the previous experiments of granting aid to individual women, which had failed.

By the 1996-97 when UNICEF phased out its involvement in the project as planned, the State government funding became a bottle neck because of a resource crunch and the scheme came to be seen as threatened. Though it had mobilized savings to the tune of Rs. 80.2 million it created anxiety and dependence and there was no autonomy.

Several other programmes showed evidence of lack of commitment to gender equality policies as for instance, default over two years in the payment of allowances that were due to rural midwives in Bihar State as also the decision to wind up the sathin programme in rural Rajasthan, turning down the demands of anganwadi workers for better pay in Karnataka.

The ICDS for providing nutritional supplements and health services to pregnant and lactating women and to the children below the age of six had hit the headlines for alleged misuse of funds.

It was the same with the Support for Employment of Women Programme [STEP] launched by the department of Women and Child Development of the government of India. Employment and income targets were not met and improvement in the status of women were not recorded.

The Central Social Welfare Board (CSWB) had socio-economic programmes for women, but the procedure intimidiated not only illiterate women but voluntary bodies also, who would like to avail the assistance or undertake schemes for the betterment of rural women.

The government approach to the dominant problems of inequalities in society, based on caste and gender disparities is typified by the Jawahar Rozar Yojna Programme in which the focus is on creating work for community asset formation (such as water tasks, mullah builders or field channels) in which the village sarpanch is the supervisor. There is no provision, which states that the sarpanch could be of either caste dominant or SC/ST or can be a woman. The case of the problem is that in this programme the weaker section SC/ST women are ignored.

The New Employment Guarantee Scheme (EGS) through Shram Shakti Voluntary community labour was introduced for women belonging to SC/ST group and below the poverty line. The latest State sponsored projects for poor women is the Mahila Samruddhi Yojna was announced as a gift to rural women on Independence day for the social elevation of rural women. Under this scheme a women who deposits Rs.300 in a post office account for a year, receive an incentive bonus of Rs.75. The approach at its best is one-dimensional empowerment. In this approach though a woman's economic situation may improve, her gender equation within the family may not change.

A Tribal Cooperative Marketing Development Federation was set up in 1987. 12.15 million SC/ST families were the beneficiaries in this context former Prime Minister Rajiv Gandhi said that out of every rupee spent for tribal welfare and betterment, only few paise actually reached the intended beneficiaries.

The CDP was lauched in 1952, with womens empowerment as one of it's aim but was unsuccessful due to various reasons.

Empowerment of Women : Role of NGOs

In implementing the programs administered by government, the NGOs come to play a very significant role. Many voluntary agencies have taken up the challenge of empowering women, some well known projects having national and one among them is international network, working solely for women's empowerment are SEWA (Self Employed Women's Association) working in Gujrat since 1971 to give full employment for women and aims for self-reliance for women economically and help them in decision-making processes.

AWARE (Action for Welfare and Awakening in Rural Environment) formed in 1975, aims at a sustainable pattern of development for rural women especially emphasising women of the dalit and tribal communities.

The vital contribution by the NGOs for women in Marathwada region studied and observed are also illustrated. Especially impact of the Adgaon project on the women is detailed.

Marathwada Sheti Sahayya Mandal (MSSM)

Marathwada Sheti Sahayya Mandal is a pioneering Voluntary Organisation working in central Maharashtra. The achievement of MSSM in watershed development have, thus far been in the area of resource regeneration and economic development. Adgaon project, a major breakthrough in the eighties in this direction, threw up some issues regarding involvement of women in development activities, processes and its advantages.

Analysis of the success story of Adgaon highlighted the gender difference, in terms of roles performed by and expected of women in the watershed. The analysis also opened up avenues of interventions towards involvement of women in the development process.

The Background of the Women of Adgaon

The socio-economic setup and general environment in the region are not in favour of women. Women face several problems as individual member of society, namely regarding education, early marriage, responsibility of stability in family organisation, say in financial matters, etc.

Educationally, most children get primary education, and many go for secondary school when the facilities are available in the village itself. Girl children are rarely sent to school outside their village. Literacy and education among women are, hence, very low compared to men.

Uncertain general environment gives feeling of insecurity for young women and a grown-up girl is often regarded as a great burden on parents. This results in early marriages, putting the burden on women at an early age. The stability in the family is normally regarded as responsibility of the women in the household. She has to fulfil this role with great care, and mostly it is she, who compromises her desires or needs in the process. Although management of household is the responsibility of woman, she has a very limited or no say in financial affairs of the family. Most decisions about family where financial investments or expenditures are concerned, are taken by the male members of the family. It is generally regarded that women's role is confined to domestic duties alone, and they cannot have a role in public life. Women rarely are seen taking part in village affairs, participation in some religious festival a few times in a year is the only public activity. Exposure to outside world is very limited among rural women. So much so that many women from several remote villages have never visited weekly market in the vicinity. This lack of exposure is also found to make them more dependents on the men.

When the Adgaon project was initiated, MSSM attempted to make the technology manageable by the local people. The achievements not withstanding, Adgaon project had shortcoming in two areas, viz, the project could not evolve, as learning, a well defined the strategy for community organisation, and secondly, women's involvement. While the efforts for community organisation were mainly ad hoc and often limited to locally solving problems as they arose, the aspect of women's involvement was largely ignored in planning and implementation of the project. With respect to women involvement, attempts were made only at a later stage in the project. Initially, women were not separately approached in the development planning, primarily because it was believed that women and children are integral part of the family institution, which gets represented by the farmer (or head of household) in public affairs. This aspect was looked into later, through discussions with several development workers and through formal reviews; but it did not help alter the programme strategy. It was only after the project was over, some meaningful analysis could be done.

Similarly, a study was commissioned by SDC in August 1998 to analyze the impact of watershed development on women. The analysis threw up several issues about gender and women's empowerment, which helped MSSM in subsequent projects to work towards women's involvement in a clearer way.

Perspective on Women's Involvement

The above observations highlighted the following aspects in relation to women's role and position in society. This understanding, although based on prevailing situation in rural areas, can be broadly generalised for the entire society and stated as premises in the following paragraphs;

- Role of women existed for centuries and any deviation is discouraged by the society, primarily due to conditioning of same over generations.
- Although most work in agriculture and almost all-domestic chores are done by women, society usually avoided recognising the importance of women's role in these spheres of life.
- Society, by and large, discouraged the involvement of women in social activities and public life, probably because women are generally believed to lack competence, and hence, not given any opportunity to prove their capabilities.
- The above factors, along with the prevailing systems of social sanctions, deprived women of any opportunities of confidence building and induced inhibitions.
- Conventional development pattern generally failed to consider women's role in planning and operationalisation of development efforts.

MSSM looked closely at the gender aspects after the evaluation of Adgaon project, which revealed inequitous distribution of benefits among men and women. The men gained a great deal more than women on technological advances and the women didnot advanced even marginally on technological front (Fig. 1). Such a finding made MSSM to relook at its interventions and strategy in order to address the issue of differential benefits.

At the organisation level, it was realised that the earlier approach of addressing the head of the household, a man, and assuming that he represents the concerns and interest of all family members, including women, was not correct. Therefore, MSSM initiated measures to address women separately and independently. In the process, some prominent factors were identified as responsible for the status of rural women and for lack of their participation in development activities. Although these were beliefs and opinions, they got continuously reinforced through repeated patterns.

- Women's role was seen mostly in the domestic sphere, and they were not believed to possess any ability to be in the social or public domain
- Women were traditionally not given an opportunity to show their talents in public or social activities, and thus were not encouraged to take part in development activities
- Women were not believed to have any knowledge of land and water or watershed, and hence, were not considered to be in a position to contribute to planning

This followed several small activities addressed to reduce the drudgery of women, viz., kitchen garden, mandals, savings and credit groups to help them meet their sundry credit requirements, as also to help them gather skills and aptitude to manage economic activities. These were useful in overcoming the initial inhibitions in working together of both, the organisation and the women. It also helped to see the limitations of working with women alone in isolation of the mainstream development work.

MSSM attempted to build upon the lesson and to focus the efforts on gender and not narrowly on women alone. In subsequent watershed projects, MSSM made some simple interventions towards both women's empowerment and integrating gender. Some of these were systematically planned while some others were not. The planning was usually done in a series of meetings with mahila mandals and watershed committees. Normally the ideas emerged during the discussions on ongoing activities and on their expansion or follow-up. Ideas developed into workable plan were discussed with the watershed committee for finalisation. This process usually took 3-4 months. Other organisations working closely with MSSM and GRASP (another voluntary organisation working on community-based natural resources management in and around Aurangabad district) followed the footsteps, and the interaction helped in refining the methods gradually. Developing the capacity of the human resources] field workers was also attempted in the process.

Major Interventions

The interventions were initiated towards awareness and skill-building and empowerment of women as concurrent strategy and a pre-requisite. Simultaneously, efforts were made in institutional field to bring together men and women and involve them meaningfully in various activities as follows:

Awareness and Skill-Development

- Specific training programme, e.g., fertiliser application, pest control, weed control, etc.

- Involvement in broad or overall village development planning through PRA
- Net planning - planning development of one's farm using simple techniques
- Gully plug construction
- Concrete mixer operator
- Women supervisors (in neighbouring district)
- Bicycle riding

Socio-economic Development

- SHGs and linking with the banks
- Tree plantation incentive scheme
- Kitchen garden
- Dal mill
- Garlic trading

Institutional Integration

- Setting up of farmer's clubs
- Training on PRI functioning
- Watershed Committees and Gram Sabha

While some of these were one-off events, the others unfolded into meaningful and useful activities.

Training on Technology

Initially, efforts were made to train the women on unusual skills or techniques not failing in traditional roles of women like concrete mixer operator and gully plug construction. However, these did not yield any significant results except that they were appreciated by the villagers. Attempts were made to train the women in supervision skills, but they never practised these skills. Thereafter, the training inputs were about the traditional roles of women like weeding and fertiliser application.

Fertiliser Application Raining was preceded by a village level seminar, wherein the women also participated and the farmers interacted with scientists on agronomic practices of certain crops. As a result women started discussing, within family and with others, on timing and costs of various operations, as also influencing the decisions.

Pest Control Training was a part of larger Integrated Pest Management Programme on cotton in Jalna district. Although the roles of scouting of insects (measurement) was with youth, the women and children were educated on the identification of insects, friends and foes, and their control. The result was that the women became familiar with the control measures and on the insecticides to be used at various stages.

Special Skills On Watershed Planning were imparted through a series of practical training sessions on net planning - a method evolved by MSSM to plan soil and water conservation treatment by involving farmer. In one village, the women and men were involved in the planning exercise, with the result that the women's participation in VWC meetings and Gram Sabhas increased considerably. Further, it demonstrated that some women have better knowledge of their farms and have a good understanding of how their farmland could be improved.

Socio-economic Development

Although efforts were made to promote income generating activities for women, the response was limited, because the identified activities fell largely outside the areas and interest familiar to them. Efforts were made to train women on nursery raising, budding and grafting, but the latter's utility remained transitory as the demand gradually reduced. So, MSSM decided to go in small steps and promoted savings and credit groups. Most groups were eventually linked to the banks, which provided the opportunity for women to interact with institutions and also to understand the economics/business tactics. They proved to be strong primary units for building mahila mandals or village committees and for launching income generating activities. A dal mill and collective trading of garlic were planned in Asarkheda and Kadwanchi watersheds then.

Likewise, kitchen garden and tree plantation was promoted through women. A scheme was designed to provide incentive to farmer for protecting and tendering the saplings on their own farms. Initially, the results were poor. But later on when the scheme was implemented through the women, there was a marked improvement. One main reason for the success was the scale - the women knew exactly how many saplings they would be able to take care of, and accordingly decided to rear only so many. This experience was useful in the Gram Sabha and committee meetings for making many other operational decisions.

Institutional Fields

This was the area where conscious efforts were made to bring together the men and women in a systematic manner. The efforts started with eliciting support from men folk and the watershed committee for women's activities. The representation of women in the watershed committee was made through SCGs and mahila mandals.

One major achievement was the farmer's club. The club membership is for husband and wife from a farming family, and not to single farmers. The members meet once every month to share and discuss their experiences and experiments on their current crops. The meeting is held in rotation on different farms - not in house. These were useful in getting the men and women together on crop production technology and the interaction was useful in raising the familiarity of technology among men and women together. It helped them understand each other's roles and intricacies of managing them. The concept did take roots in only one village out of seven in which it was launched.

Dilasa

This organisation has a range of activities for women development. It has a multipurpose women centre. The objectives of this centre is to give legal aid, family counseling, income generating activities and rural enterpreneurship. This organisation has formed 22 womens co-operative dairies. The rainwater harvesting structures have reduced the womens burden in fetching potable water. With the help of women and women masons 104 structures have been constructed to store water. This is an unique feature as here women have been trained as masons. For the emancipation of women the project of smokeless chulla were made.

The Development of Women and Child in Rural Areas (DWCRA) programme have been implemented. through Dilasa. This programme focuses on women below the poverty line to give them opportunities for self-help and employment. National Institute of Rural Development has endorsed this programme as remarkable.

Abhinav Vikas Sanstha

This organisation has been tackling the problem of womens low literacy rate and exploitation of women. The organisation developed self-help groups for women among in which many members made savings of 4 lacs.

Nirman

This organisation started self-help groups and mahila mandals to strengthen the women. It also organises in house training, special training for leadership development, help, education, gender development and other trainings to empower the women of the area.

Manavlok

The special women oriented project of this organisation is Manaswini' which means one who propogates her own enlightment with grit.'

The rural women especially the poor are organised to form Bhumikanyamandals' in which the members come together to discuss, solve their problems and join Krushak Panchayats in development programmes. It also lends legal advice and support for destitute and deserted women.

Janarth

This organisation has awareness programme for women to build confidence in them by increasing their knowledge base. The various programmes undertaken are formation of mahila mandals, incentive schemes for participation in mandal activities are also included like, seed gathering, medicinal, nursery raising, collecting of medicinal raw material. This organisation also conducts workshops for women and girls on their legal rights.

Grasp

This organisation promoted 21 self-help groups. It also works for gender. sensitization focussing on empowerment of women and enhancing their participation in the developmental process. The organisation ensures womens participation as stake-holders in the management of watershed development. They also promoted saving groups for women which followed into savings and credit and then to enterpreneurial activities and community banks. This organisation understood the importance of women participation in development and they empowered women by increasing their participation in decision-making levels and take up income generating activities. They provided them with infrastructural support and interacting with women through one to one basis and through the community.

Addressing the gender problem, it is learnt from the working and objectives of the various voluntary agencies studied and visited, that the women's issues have to be approached in a multipronged way. It is also observed that if one approach is not found to bring about positive changes the agency had to make many necessary changes and develop innovative schemes to suit the local women. It was also observed that once the agency managed to hold the interest of the women, the women got involved fully and worked tirelessly to make the programme effective and sustainable for themselves and their village.

References

V. Mohini Giri : Forward - Empowering Women; An Alternative Strategy from Rural India in Sakuntla Narasimhan, Sage Publications, New Delhi, 1998, p. 10.

Bina Roy : The Status and Role of Women in Our Changing Society - Social Development Essay's in Honour of Smt. Durgabai Deshmukh, edited by B.N. Ganguli, Sterling Publication Pvt. Ltd., pp. 58-59.

Boutros Boutros Ghali : Secretary General of the UN, in his welcome speech on the Fourth Congress on women in Beijing.

Amartya Sen : In The Hindu] Delhi , 6 November 1995. Interviewed by Rammanohar Reddy.

Batliwala S. (1993) : Empowerment of Women in South Asia : Concepts and Practices Sponsored by the Asian South Pacific Bureau of Adult Education and FAO's Freedom for Hunger Campaign, Action for Development, Banglore.

Pillai J.K. : Women and Empowerment] (ed) 1995, Gyan Publishing House, New Delhi, pp. 38-40.

Bhasin (1992) in K.C. Vashista's and Sashi Malik's : Some Strategic Effort Towards the Empowerment of Women, Conceptual Framework of Women's Empowerment; University News, Vol 4, No. 5, Feb. 4-10, 2002, pp. 12-16.

National Consultation on Women Executives in Cooperatives Gender Discrimination in Work Places, Conducted by Dr. Medha Dubashi in Collaboration with ILOCOOPNET April 1998 .

Longwe S. 1990 from Welfare to Empowerment, a Post UN Women's Decade Update and Future, Directions, Working Paper 204, Michigan State University.

Government of India (GoI) 1998 : National Perspective Plan for Women (NPPW) Department of Women and Child Development, Government of India, New Delhi.

Gro Halem Brundtland, Prime Minister of Norway at the Fourth World Congress on Women 1995 - Beijing.

Country Report of Government of India, 1995 : Fourth World Conference on Women, Beijing, 1995, Department of Women and Child Development, Ministry of Human Resource Development, Delhi, p. 45.

Role of NGOs in Making Pregnancy Safer : Prof. S.N. Mukherjee, World Congress of Women's Health, Calcutta, 2000.

Women's Access to National Resources : The Need for Advocacy for Grassroot Demonstration - Presented by Anjali Vijay Kulkarni in the 8th National Conference : Survival and Sovereignly : Challenges to Women, Studies by Jaws held at Pune, May-June 1998.

Indian express.com, July 31, 2015

Editorial, The Times of India (New Delhi), 26 December 1992.

India : The Road to Human Development, UNDP, 1997, New Delhi, p. 7.

Estimates of the Extent of Poverty Range from 44 to 70 Percent, According to Different Studies and Definitions of Poverty.

Sakuntala Narasimhan : An Overview of Past State Initiatives, Empowering Women (ed), Sage Publications, New Delhi, pp. 36-37.

Ramola Baxamusa and Hema Subramaniam, 1992 : Assistance for Women's Development from National Agencies : Employment Programmes, Mumbai, Popular Prakashan, p. 70.

Newstime] Hyderabad , 5 July 1996.

Ramola Baxamusa and Hema Subramaniam, 1992, op.cit.

SEWA - Source Internet.

Sakuntala Narsimhan : Empowering Women - Alternative Strategy from Rural India, Sage Publications, New Delhi, 1999, p. 134.

Yugandhar Mandavkar : Case Study on Gender and Watershed Management, Paper Presented, Gender and Technology Workshop, organised by Anthra at Hyderabad on July 27-30, 2001.

CHAPTER-7

Government and Voluntary Agencies

This chapter studies the nature of the government and NGO relationship, depicting the approach, control and supervision government has. The role NGOs play in relation to the government has been described while studying various NGOs in Marathwada.

It is generally agreed that the development strategies and programmes implemented during the last three decades have failed to effectively tackle the causes of rural poverty. The benefits of whatever development and growth has taken place have not trickled down. Landlessness has increased and so has poverty, unemployment and inequality. Peasants, landless people, the small fisher folk, plantation workers, roadside laborers and women have been marginalised. In addition to this through environmental destruction, the very resource base of the people is being rapidly destroyed. In most South Asian countries, some local and national non-governmental organisations (NGO's) have been actively involved in experimenting with innovative approaches to fight exploitation and to initiate a process of participatory development.

Since there are less indications that any transformation will occur, the capital intensive industrialization and the resulting urbanisation will continue to spread, the natural resources will be further depleted and the rural poor, on whom such policies have a disastrous effect, will be impoverished and marginalised. The urgency of the situation for the rural poor cannot be over emphasized because in effect the policies have virtually resulted in a war on rural poor. The rural poor, who form a majority in these countries, must therefore be empowered to regain the ownership and control of resources, like land, water and forests. Thus, there is an urgent need for organisations, which can work towards the empowerment of the rural poor and for an alternative vision and process of development. Atleast some NGO's could play this role of an alternative model of development and on a common program of mutual linkages, support and collaboration.

Intervention of Voluntary Agencies

India has a longest history and the widest experience of NGO's working with the rural poor. Prior to 1947 and the Independence Movement, NGO's encouraged the development of handicrafts, participated in the struggles for social reform and for the emancipation of the oppressed. In the 1950s the emphasis shifted to asking for restoring land and other resources to the rural poor. Both Gandhian and Marxist groups participated in movements related to development of the poor and this gave impetus to a fairly widespread NGO moment's encompassing various rural sectors. Many of these focussed on empowering the rural poor but they were usually coopted or suppressed. By the 1960's most of the indigenous movements as for handicrafts sector had been taken over by the state, by then the NGO's realised that the poverty situation could not be reversed by working within the system and working outside it subjected NGO's to government control and harassment. This dilemma remains unresolved and is infact filled by a contradiction that while effective NGO's are suppressed, government is increasingly recognising their importance and allocating more resources for their work The 7th five-year development plan for instance has a separate chapter an NGO's. But some of the NGO's felt that this move is more towards the control of NGO activity than for encouragement of it because by funneling resources though it through own channels and by controlling foreign funds to NGO's the government can oversee NGO activity and direct it towards its own purposes.

Government and Voluntary Agency Relationship

Government and NGO relationship is supportive as well as controversial in many areas. It can be understood in the following observations.

Role identification at the NGO level starts with the realisation that the task is tremendous and the government has limitations in terms of its orientation, approach, resources, manpower and local level support system among the beneficiaries and other constituents in the community.While the NGOs generally enjoy a state of open collaboration with the government in developed countries, many of the NGOs which are operating in India learn to live in a situation of distrust and suspicion with the government machinery, particularly at the local level.

Alternative measures adopted by NGOs to address a problem are often looked down upon by he government, particularly the bureaucracy at the local level. Mutual cooperation and support between NGOs and local level bureaucracy depends mostly on the perspective of the individual bureaucrat rather than government norm or policy implementation. While there are limitations to institutionalizing expected behavior of a government officer due to local level variation and differing characters of NGOs operating in the area, individualization of development perspective is found to be useful as it is based on personal orientation and involvement of the bureaucrat in NGO work. Personal factors like style of functioning, past experience, interpersonal relationship with the NGO leaders etc. play a critical role in improving the efficacy in relationship between government officials and NGOs.

The question of funds flowing from the government to NGOs is an important criterion in shaping the relationship between the two. There are two distinct opinions on this issue. One group suggests that NGOs should function independently. They should neither receive government funds nor directly implement government programs as it affects independence and voluntarism. The government policy is in favour of supporting and directly funding NGOs. But many NGOs perceive this as the beginning of government control of the NGO sector. In support of this point, NGOs recall the experience of the cooperative sector. Provision of government funds for the cooperatives resulted in bureaucratic control and political interference.

The other group propagates the idea of cooperation of NGOs with the government to achieve the common goal of improving the conditions of the poor. None of them can and should really function independent of the other to fulfill their objectives. It is felt that role of educating, organising and reaching out to the poor which is essential in any program, cannot be played by the local bureaucracy. Since the government is in a position to help NGOs with financial assistance, NGOs may receive funds directly from the government in achieving their objectives. There are supporters for both the groups among the bureaucrats, politicians and NGOs. There is another group in the NGO sector, which believes in a middle path. According to this moderate group, NGOs need not receive government funds directly, but they may mobilise the target group to derive benefits of the government's anti-poverty programs directly. These NGOs support the implementation of government programs without directly receiving and accounting for government funds. This group sees the NGOs as catalytic agents.

Since Independence, several government departments and ministries have been turning to the NGOs for assistance in designing and executing training and extension programs. As early as 1955, the Ministry of Education selected certain NGOs to design, develop and promote rural higher education according to the changing needs of the rural areas. The Ministry of Social Welfare, Ministry of Health and Family Welfare, and the Ministry of Rural Development are the three other crucial ministries, which have been independently providing financial assistance to the NGOs.

The concerned ministries responded to the NGOs in two ways: a) By providing funds for the government designed schemes to be executed by the NGOs, and b) By providing funds for theNGO designed schemes to be executed by the NGOs themselves. Considering the importance of NGO involvement, the ministries, in the recent past, have started making some efforts to draw the officials who are interested in working with NGOs, to staff the concerned sections for maintaining a healthy relationship with the NGO sector.

After Independence, the Government of India identified certain areas for State intervention such as health, education etc., and assumed limited responsibilities in sectors like social welfare. The government provided support to NGOs to undertake projects for the benefit of women, children, youth, differently-abled and destitute through the Ministry of Social Welfare at the Central and State level and through government sponsored autonomous bodies.

Five Year Plan and Voluntary Sector

The voluntary sector has been given due importance in the planning process right from the First Five Year Plan, as emphasis was given on public cooperation in national development with the help of NGOs. It was highlighted in the First Plan document that the "Public cooperation and public opinion constitute the principal force and sanction behind planning. A democracy working for social ends has to base itself on the willing assent of the people and not the coercive power of the State." In the Second Plan, it was reiterated that public cooperation and public opinion constitute the principal force and sanction behind India's approach to planning. It was observed that wherever the people, especially in rural areas, have been approached, they have responded with eagerness. In national extension and community project areas, in local development works, in shramdan, in social welfare extension projects and in the work of voluntary organisations, there has always been willingness and enthusiasm on the part of the people to contribute in labour and local resources have been made freely available.

The Third Five Year Plan emphasized that "The concept of public cooperation is related to the much larger sphere of voluntary action in which the initiative and organisational responsibility rest completely with the people and their leaders, and does not rely on legal sanctions or the power of the State for achieving its aims. It was realized that so vast are the unsatisfied needs of the people that all the investments in the public and private sectors together can only make a limited provision for them. Properly organised voluntary effort may go for towards augmenting the facilities available to the community for helping the weakest to a somewhat better life. The wherewithal for this has to come from time, energy and other resources of millions of people for whom NGOs can find constructive channels suited to the varying conditions in the country." During the Fourth and Fifth Plan, the thrust on public cooperation and involvement of people's organisation was lost due to attack on over territory and recession that followed. During this period investment was focussed especially in intensive agricultural programmes.

In the Sixth Five-Year Plan, the idea of participation of people's organisations was again recognised. Success stories in the field, of NGOs like the IHMP's Project on child and health care in Maharashtra, Bharat Agro Industries Foundation's work in animal husbandry and social forestry and Self-Employed Women's Association (SEWA) were quoted and it was stated that the country is dotted with numerous examples of highly successful voluntary action of this nature.

Role of NGOs in development got a further fillip in the Seventh Five Year Plan where it was declared that serious efforts would be made to involve NGOs in various development programs to supplement the government efforts to offer the rural poor choices and alternatives. The emphasis continued till the ongoing Ninth Plan, wherein efforts are being made to promote peoples' participatory bodies like Panchayati Raj Institutions (PRIs), Self-help Groups and NGOs for development. The following criteria were identified for identifying voluntary agencies for enlisting help in relation to the rural development programs:

- The organisation should be a legal entity
- It should be based in a rural area and be working there for a minimum of three years
- It should have broad-based objectives serving the social and economic needs of the community as a whole and mainly the weaker sections. It must not work for profit but on 'no profit and no loss basis'
- Its activities should be open to all citizens of India irrespective of religion, caste, creed, sex or race
- It should have the necessary flexibility, professional competence and organisational skills to implement programs
- Its office bearers should not be elected members of any political party
- It declares that it will adopt constitutional and non-violent means for rural development purposes
- It is committed to secular and democratic concepts and methods of functioning

In the Eight Plan Document, due emphasis was given on building up people's institutions. It was admitted that developmental activities undertaken with people's active participation have a greater chance of success and can also be more cost-effective as compared to the development activities undertaken by the Government where people become passive observers. It was admitted that a lot in the area of education (especially literacy), health, family planning, land improvement, efficient land use, minor irrigation, watershed management, recovery of wastelands, a forestation, animal husbandry, dairy, fisheries and sericulture etc. could be achieved by creating people's institutions accountable to the community. Therefore the focus of attention will be on developing multiple institutional options for improving the delivery systems by using the vast potential of the voluntary sector.

In the Ninth Five Year Plan, it was admitted that private initiative, whether individual collective or community based, forms the essence of the development strategy articulated in the Plan and efforts to be made to eliminate disadvantages which had prevented some segment of our society in participating effectively in the development process. Keeping up with this line of thinking, "promoting and developing people's participatory bodies like Panchayati Raj Institutions, cooperatives and self-help groups" was one of the objectives of the Ninth Plan.

The Mid-Term Review (MTR) of the Ninth five-year Plan (October 2000) has also documented some successful and sustainable projects undertaken by NGOs and other people's organisations. The MTR has identified following problems in the performance of central and state plans. Some of these maladies faced by the country can be mitigated by the emerging voluntary sector as a compliment and supplement to the state efforts.

- The significant factor can be effective governance to ensure balanced growth of the different regions of India.

- Development is an outcome of efficient institutions rather than the other way around. There is a need to focus on building institutions and improving governance from maximizing the quantity of development funding to maximizing of development outcomes and effectiveness and efficiency of public service delivery."

Further, under the 'Directions for Reforms' of MTR, it has been inter-alia suggested that "Initiatives by local bodies, NGOs and women need to be encouraged", to resolve some of the above-referred problems.

It has been observed in the Approach Paper to the Tenth Plan that in many States, we have: hospitals / dispensaries but absence of personnel and there are school buildings but teachers remain absent. To rectify these anomalies and to achieve most of the targets set up for the Tenth Plan, the need to promote voluntary sector has been recognised. The theme of encouragement to voluntary sector continued in the Approach Paper to the Tenth Plan and reflected in the following words, "In view of the continued importance of public action in our development process, increasing the efficiency of public interventions must also take high priority". The minimum agenda proposed in the Approach Paper has recognised voluntary sector by putting "greater decentralization to PRIs and other people's organisations" as one of the items for the Tenth Plan.

The voluntary sector deserves encouragement to ameliorate at least some of the above-referred problems, because of its following comparative advantages;

- NGOs are much closer to the poorer and disadvantaged sections of the society
- Their personnel is normally highly motivated and altruistic in their behaviour
- They can easily stimulate and mobilize community resources and have access to volunteers
- They are more effective in bringing people's participation
- They are less rule-bound and are non-bureaucratic, non-formal and flexible in their structure and operations
- They have greater potential for innovations
- They prefer to work in a multi-sectoral framework
- They are catalysts for creating social cohesion

Certain disadvantages or shortcomings of voluntary sector are also well-known, namely, their inability to cooperate with each other in a way which would allow for coherent policy making, their accountability and transparency is not perfect and their operations are smaller in scale. Therefore, there is a need to improvise the working of NGO's by scaling up their operations and by making them transparent and accountable.

In examining the government-NGO interface in India, it becomes significant to note the process through which the government has been directing and controlling the contribution of NGOs (for example, through funding literacy work, Krishi

Vigyan Kendras for agriculture, work on differenlty-abled, etc.) in the country's development process . From the very first planned effort, the government has been explicitly stating the significance of NGOs in the country's development efforts and also planning for the NGO sector. At times, the government officials, who are engaged in preparing the blueprint of the programmes to be executed by the NGO sector, have no experience in working with NGOs. As a consequence, only large NGOs with their bureaucratic pattern of functioning have been in a position to receive government money. Small NGOs working in inaccessible pockets cannot reach out the benefits of such government schemes to the most needy sections of the rural population.

The concept of NGO, the importance given to the NGOs and their contribution/ role expectation of the government in development efforts have undergone a significant change over different five-year Plans. During the period of seven Five Year Plans, the government perspective towards the NGO sector changed from government planning for the NGOs and funding schemes to government encouraging the NGOs to plan for themselves and advancing financial assistance to them. This is a very positive trend.

The Indian State has been initiating and implementing various welfare schemes for poor and the underprivileged. The Governments policy for liberalisation of the economy and India's stepping into the World Trade Organisation (WTO), was contemporary with the announcement of reservation for backward classes, the constitutional status for the local governments and welfare schemes for the poor. The Indian government also started popularizing the concepts of empowerment, community participation and developmental changes. Especially, the involvement of the NGOs and the community participation became handy words among the policy-makers dealing with poverty alleviation .

Importance of Voluntary Agencies

The dependence of the State on the NGOs for the implementation of government sponsored programs is recent. The Indian State confidently implemented development policies without any internal support. In the fifties the government initiated Community Development Program, Panchayat Raj system and the Cooperative Movement to encourage community participation. The NGOs were then taking up welfare activities like education, health and rural development. Most of these were Christian or Gandhian organisations. The State had no problem in dealing with the NGOs, as they never questioned the system nor did they make any effort to understand the roots of the social problems.

In the mid-sixties, the government confidence was shattered with the economic and political crises. The failure of Community Development Program, the Panchayat Raj system and the Cooperate Movement led to bureaucratization and the monopolization of political and economic power by the rich and the neo-rich communities. This affected the legitimacy of the State. Serving this phase many NGOs emerged. They upheld the rights of the marginalised communities and poor masses. The NGOs took up some programs of life, strategies in rural development, housing, community health, and environment. This brought the NGOs in direct conflict of the Indian State, as there was the shift from welfare to development activities. The State began to look at the NGOs as competitors. The government, on one hand announced several poverty alleviation programs, and on the other hand enacted laws, to regulate and control the NGO activities. The NGOs, which refused to follow the government rules, were declared illegal and even anti-national. Many NGOs criticized the State intervention in their activities.

The government was implementing many schemes for the poor, but the benefits did not reach the actual beneficiaries. The then, Prime Minister Rajiv Gandhi admitted that only 15 per cent of the funds reached the poor. This brought a great discontentment among the poor sections, which came up with a spur of autonomous struggles of the dalits, women, tribal, farmers and workers. The government instead of trying to resolve the problems was more for the policy of globalization and liberalisation, which burdened poor sections. The people could see with naked eyes the policies of government were anti-people. The economic and industrial state always tries to fulfill their political needs by giving a human face to the on-going economic reforms, as they cannot totally ignore the people's concern and remain insensitive to their aspirations. Since the sixth plan there has been a remarkable change in the attitude of both the government and the NGOs. The Indian policy makers realized the limitations of the government and the comparative advantages of the NGOs. They understood that the government has failed to reach the intended beneficiaries through all their welfare and development activities caused by conception, red-tapism and bureaucratic hurdles.

Where the Government failed with their granted bureaucratic machinery, there the NGOs because of their easy accessibility to the poor, their awareness of local problems, their lo- cost alternative technologies and their participative strategies could make better impact in those areas. Revealing these facts, the policy makers realised the advantage of collaborating with the NGOs. Accordingly, the Indian State began encouraging the participation of the NGOs in different government sponsored welfare and development programs. The provision of grant-in-aid has been made in different government departments, inviting the NGOs to involve themselves in planning, capacity building, implementation and monitoring of the development programs. The NGOs, which have become professional and pragmatic over the years, responded positively to the invitation of the State in return for administrative and financial support. Thus the strategy of implementing poverty alleviation programs through the NGOs and the community based organisations (CBOs) gained the recognition. The Government itself began taking the initiative in identifying competent NGOs and forming the CBOs at the grassroots level. This strategy was first initiated in the rural development programs.

The crucial role of NGOs vis-a-vis government's efforts to improve the conditions of the poor and the disadvantaged has been debated since independence. In the absence of a clear NGO policy of the government, this study makes an effort to explore the role of government in the promotion of the NGO sector in India. Although government documents and government appointed committee reports reflect the need for increased NGO participation in government sponsored development efforts, only a few concrete steps have been taken in this direction.

There is an ever increasing realisation about the importance of a continuous dialogue between different levels of bureaucracy and the NGO sector to develop confidence and accept each other as equal partners for the development of the poor, deprived and disadvantaged rural communities.

Given the unique Indian situation - of geographical vastness and socio-cultural diversity within and between the states and regions, it is imperative that the government machinery will play the dominant role in improving the condition of millions of the people. Government machinery includes the vast bureaucratic establishment, judiciary, people's institutions, and government promoted development agencies. Nevertheless, NGOs will continue to have a crucial role in the overall development efforts in the country. On the one hand governmental endeavor will dominate the national and regional scene, on the other, importance of NGOs in the micro/local level intervention cannot be ignored. There are areas where NGOs and government agencies can and should work together in the interest of the people . It is a question of who plays what role at what level. Government agencies will implement programmes, which are planned at macro-level for the benefit of large sections of population spread over every corner of the country whereas NGOs will work in isolated pockets, and will work with the issues touching the people at the grassroots.

Role of Voluntary Agencies

NGO's role may be broadly classified into the following categories: First, NGOs make adaptations in applications of schemes planned by the government at macro level to specific situations considering local requirements. Often, the government, under certain funding provisions through different ministeries and autonomous agencies (like CA-PART), provides broad guidelines for proposing a project and NGOs are called upon to formulate a project proposal taking local situations into consideration. Second, NGOs undertake a role supporting the actions/programmes carried out by government agencies. Although they may directly aid to implement government efforts, these NGOs may not necessarily have government support or recognition. For example, an NGO which is engaged in conscientising the rural poor, may assist the target group to avail benefits under the government antipoverty programmes and in turn, aid the Block Development Officer in executing the Integrated Rural Development Programme. In spite of such a role the NGO may be in conflict with the local bureaucracy.

Third, the role of the NGOs may be in the form of providing approaches alternative to government policies and programmes, addressing different problems of the target groups. Ashish Gram Rachana Trust of Pachod, Maharashtra has developed a community health approach, which is considered to be an alternative to primary health care system of the government. Fourth, NGOs develop backward and forward linkages for enhancing the effectiveness of their own and of the government in undertaking development opportunities. For example, PRADAN, an NGO promoting rural enterprises as a strategy for development of the rural poor, carried out projects in Rajasthan, Bihar, Madhya Pradesh and Uttar Pradesh, involving backward (for example, exploration of opportunities and resources, identification and training of manpower, developing channels for supply of raw material and other resources) and forward linkages (e.g., promoting networking for marketing products). On the other hand, government rural development efforts have been to provide mostly credit support through IRDP or DWCRA for self-employment or entrepreneurship in the informal sector. Fifth, NGOs supplement government efforts particularly in reaching out to the less accessible target groups and at times complement existing services in response to the other needs of the same target groups.

Sixth, the role relates to more radical NGOs, which take a completely different perspective. They believe that social transformation is a necessity to change the conditions of the deprived and disadvantaged. A structural readjustment if not change, is required to transform the exploitative social system. The government, which operates in the existing structural adjustments, cannot afford to do anything, which will shake its own base. As a result pro-establishment oriented government efforts cannot be responsive to the needs and problems of the common people in the real sense. These NGOs adopted a critical role within the existing legal-political system . Across the NGOs working at the grassroots level, there is some sort of consensus on the role of NGOs in critically evaluating government policies and programmes affecting the disadvantaged. Whenever the move of the government machinery - administrative, legal or political - affects the interests of the poor, particularly the rural poor, and favors the elite, even in the rural areas, NGOs have a basic commitment to play an active role.They may even take the extreme step to conscientise and mobilise people to raise their voice in protest against the government action and to protect their own interests.

Government-NGOs Interface in India

There are many instances where NGOs are already involved in Government projects e.g. as sources of information on local conditions or for implementation of specific project components. In broad terms, four forms of Government-NGO cooperation are conceived: source of information, consultants, executing or cooperating agencies and confinanciers. The principal NGO roles in Government sponsored poverty alleviation projects in operational terms may be summarized as follows;

NGOs in Financial Interventions: The NGOs can provide, exclusively or in combination, financial intermediation services e.g. on lending loan funds to the beneficiaries, credit support services to the banks (e.g. loan appraisal, supervision, recovery) and to the beneficiaries (e.g. completing loan applications, raising awareness, group formation and strengthening).

NGOs in Grassroots Organization and **Support Services:** NGOs can support grassroot mobilization, natural resource management, reforestation/afforestation, social forestry, community organizing and training, formation demand services to village organizations, ascertain beneficiary needs, help them articulate it and organize them around sustainable grassroots economic and ecological activities, health, education and other social services.

NGOs in Social and Women's Development: They help in Mainstreaming women's participation in develpoment and addressing their social issues.

NGOs in Resource Conservation & Natural Resource Management: They can assist in policy formulation, project design and preparation of resource conservation strategies (e.g. National Environment Management Action Plan).

NGOs in Disaster Management: NGOs can support disaster relief and mitigation, preparedness and management.

To Supplement Government and not Compete: Government and NGOs objective is similar, to offer choices and alternatives to the rural poor. Government cannot opt for a monopolistic approach in rural development, so NGOs can be looked upon as non-profit agencies or private sectors in village to supplement the government effort.

To be Voice of the people at Village Level: There is a need for an independent agency to act as a reliable feed back so that government policies and programmes could be designed taking real community problems into account.

To set an example: It should be possible for the voluntary agency with limited resources to reach a larger number with fewer overheads and with greater community involvement. There are many examples, viz., Comprehensive Rural Health Programme (CRHP) in Jamkhed (Maharashtra) using illiterate and semi-literate women for preventive health programme; SEWA (Self-Employed Women's Association, Gujarat) mobilising and organising slum and pavement based female entrepreneurs in the so called unorganised sector; Marathwada Sheti Sahayya Mandal (MSSM, Aurangabad) is a NGO working in Central Maharashtra for rural development in general and watershed development in particular; Kerala Gandhi Smarak Nidhi on intensive paddy cultivation by engaging 'barefoot' technicians' which the State government adopted in the Fourth Plan period; Eklavya (Madhya Pradesh) in training teachers in science education through government schools; and MYRADA (Karnataka) in wasteland development in their own way are all setting an example. They all offer practice-oriented proposals and have policy implication when it comes to replication.

To Activise the System and Make it Respond: The NGOs can take up the tasks which are either not being observed or effectively enforced by the government. Eg; Minimum Wages Act, Abolition of Bonded Labour Act.

To Disseminate Information: Very often, schemes, programmes and projects of the government, with its many policy changes, orders and circulars are not percolated fast enough to the village level. More often, the interpretations of these schemes are left with the lower hierarchy of government functionaries, who are not wholly sympathetic to the problems of rural poor. The role of NGOs is to distribute this information strength to the rural poor through as many channels as possible, so that they can decide which scheme they would prefer for their betterment.

To Utilize Indigenous Resources for Development: NGOs can utilise the human resources, rural skills and local knowledge in the villages. This credibility of idea has already been proved in the voluntary sector, and it is time to get it accepted in the government circles.

To Make Communities Self-Reliant: Unfortunately, the development and delivery system as it exists today is designed to make communities more dependent than independent. The system of accountability within the government is so severe and inflexible that the subsidies, loans, inputs, services and support mechanisms, ostensibly designed to serve families living below the poverty line, actually tends to enscare them rather than allowing greater freedom for their work.

To Train a Cadre of Grass-Root Workers: The technology base that we think exists at the village level is virtually non-existent in practice. The demands of the community as well as the high expections have made it necessary to bring professional expertise to the poor without intermediaries.

To Mobilise Financial Resources within Community: If the ultimate idea is to make communities independent , to promote self-reliance, to generate awareness and develop good human beings, then much depends on how much value we give to self-respect and dignity. Anything given free is not appreciated.

To Mobilise and Organise the Poor: The NGOs can mobilise community to demand quality service and impose a community system of accountability on the performance of grassroot government functionaries.

There is many area of professional services for which the government does not always have the primary responsibility, professions like law, medicine, engineering, teaching, journalism, accountancy, management etc., have their social responsibilities and professional accountability., discharging of which demands public vigilance and social action. In recent times there is a steady decline in the quality of professional services available to the common man and the ethical commitments of professionals are diluted sometimes beyond tolerable limits.

Social action groups and voluntary organisation have a key role to play in vigilance and prompt action of correction or information concerning professionals. It is not adversary relationship with the government or establishment. It however should be emphasised that role of the NGO's is not to be equated with that of government functionaries, but should be prepared to question and correct government action. In order to maintain its legitimacy and credibility among the people which incidentally are its greatest assets, the NGOs must have perception of social justice under the constitution, a certain degree of expertise in the activity they propose to undertake and a character which demonstrates sensitivity, objectivity, efficiency and commitment to human rights. The responsibility of the organisations will be also to work only on the basis of mechanism that ensures standard quality and performance from every member of the group. The Government-NGO collaboration has been working increasingly in the areas of a micro-credit income and employment generation, education, health environment, population, women and youth development livestock, fisheries, resettlements disable management etc.

A review of the collaboration indicates three major types of arrangements: (a) Sub-contract; (b) Joint implementation; and (c) Government as financier of NGO projects.

Voluntary Agency and Government

The most common collaboration is the sub-contracting arrangement where Government agencies enter into contracts with NGOs.

Joint implementation on a partnership arrangement, where NGOs are involved either as co-financier or joint executing agency with the Government, is least practiced. In the area of micro-credit there is an emerging trend for the Government to finance NGOs credit operations.The development of a sustainable collaboration and partnership requires the acceptance of some fundamental proposition by both the Government and NGOs.

PRIs and NGOs

The 73rd amendment in Indian constitution has drastically changed the nature of Panchayat Raj (institutions of village self-government). They have become a third tier of government. However, this move of democracy at grassroots level, despite many strengths has certain degree of vulnerability and some weaknesses. NGOS who work towards the same broader goal as Panchayat, have also become apprehensive about all tiers of government.

Strengthening PRI means making PRI effective enough to deliver what it is supposed to deliver. PRI had so far worked as a democratically controlled, State-funded delivery agent of government, with a semblance of democratic set up. The amendment makes it clear that PRI is to function as an institution of self-government and discharge responsibilities related to development. PRI has to work as an institution of self-government therefore governance at the local level is one of its major tasks. The second important task for PRI is to act as welfare government i.e. to initiate, plan and execute the economic development activities and to effect justice. The amendment makes it mandatory to constitute a District Planning Committee (DPC). The PRI has to act as the executor of development schemes entrusted to it and implement them professionally. All PRIs now not only have an operating structure but many constitutional support bodies as well as the State Election Commission, State Finance Commission (SFC) and District Planning Commission (DPC). PRI being local and democratic has to be amenable and responsive.

Inspite of the strengths the PRI's has acquired after the 73rd amendments there are some weaknesses as follows;

- The biggest weakness of PRI is that the State government legislates on its autonomy and responsibilities which makes it ineffective.
- Secondly, though PRI has autonomous organisational structures, they do not have the requisite manpower.
- Thirdly, in addition to the lack of stable and sufficient autonomy, PRI is poor with its financial resources.
- The State government has supervisory power over PRI; and
- The another important weakness of PRI is the lack of trained manpower to discharge its responsibilities.

Strategies for Strengthening PRIs through NGOS

In a democratic set up with strong market forces, a large number of people are left in the margin. In the market economy a large number of people get marginalised because they lack capital and other resources. NGOs work for a civil society where these marginalised people and groups can relate themselves with organised democratic process and take control of development activities and their own surroundings. PRI can be a government of these marginal groups. The authoritative hold over local policy, resources and priorities by people at grassroots level is what most NGOs endeavor to promote. Both PRI and NGO would make rural masses self-respecting and confident with power of self-assertion and dissent. Therefore, it is imperative that NGO works to strengthen the capacity of PRI to deliver its own objectives.

NGO can seek help of PRI to get those things done, which need authoritative power. NGOs possess certain technical and professional resources, which are indispensable for effective development work at grassroots level, and PRI is going to need them. Many NGOs got a strong backing of the government for their good work and there is no reason PRI will not appreciate and solicit their valuable help. Infact, being the local body, it is more liable that PRI can facilitate its work as the beneficiaries of NGO endeavors are going to be the local people.

Strategic Plans for NGOs in PRIs

NGO can play multiple roles for all the three major objectives simultaneously. At a micro-level NGO can help PRI in arranging its capacity for delivery of development projects, good governance of public and articulate the PRI demands for greater autonomy and resource allocation. At micro-level NGOs have to help the deprived sections to contest PRI elections seriously. NGO has to help people to send genuine representatives of reserved categories, not the one who are product of dominant section. The other task were be to add and assist lakhs of representatives to do their job well by increasing their technical and professional competence. NGO can perform a crucial job of training these people at grassroots level in the process of governance, in their legitimate rights and say in PRI for shaping the agenda, policy and priority of PRI at grassroots levels.

NGOs are most suitable to train the women and other first time entrants to PRI at Gram Sabha level. This training can be for a variety of purposes e.g. in participatory decision making, conduct of meetings, procedures and process for different PRI activities, details of different government programmes entrusted to PRI etc. and for Zilla Parishad, the training in governance process, aggregating the local - level plans and preparation of district plans and above all in the implementation of development projects, professional assistance of NGO can play a crucial role.

PRI with professional and technical support of NGO can show dramatically better results. To effect such results, PRI needs NGO because the bureaucracy and established interests join hands not to allow the benefit of programmes to reach target groups.

If the NGOs cooperate with PRI and make effective use of the huge development money coming to PRI, the credibility of PRI as competent and dependable organisation will increase. This will also increase people's faith in PRIs delivery capacity and its credibility, as development organisation will rise up in the society. However, NGO must play their role in such a way that over a period of time their role becomes reluctant. They must not become more attached to PRI. Having provided

PRI with expertise, confidence and credibility it must gradually withdraw from these areas. Besides enhancing the credibility and delivery capacity of PRI, NGO needs to have long term action plans. There is need to make a concerted public compaign for promotion of PRI, which can counter the propaganda and tactics of PRI detractors who brand it as a dead horse.

Government-NGO Collaboration: Problems and Suggested Measures

- The first problem is communication gap in identifying respective jurisdiction, the suggestion given is creation of a permanent platform at national and local level for mutual interaction to make collaboration operational and effective.
- The next problem is lack of trust, the remedy is that conducive policy and laws should be formulated and more joint collaboration projects should be launched.
- The problem of lack of recognition of potential is overcome by formulating rules and institutional environment facilitating healthy and effective growth of NGO activities to eradicate poverty (particularly micro-credit operation).
- Strengthening institutional capabilities to promote collaboration can solve the negative attitude of respective employees towards each other.
- The problem of delayed approval procedure is overcome when implementers given the flexibility to take necessary steps for minor adjustments during implementation.
- The problem of insufficient accountability is looked upon by ensuring the transparency of NGOs.
- The lack of workable partnership contract is surmounted by a written contract among partners and NGOs should be considered as development partners.
- Inadequate co-operation and co-ordination from the government is overcome by frequent meetings and consultations with NGOs, better co-ordination and spirit of partnership should be developed.
- Corruption and bureaucratic harassment can be solved by direct access to concerned authority to avoid corruption red-tapism, greater mutual understanding and improvement in management.
- Irregularities and delays in fund flows can be solved by ensuring regularity and certainity of fund availability.
- Inadequate technical or skill and institutional and logistic support is corrected by providing skill development training for the relevant personnel . NGO representatives should be included to resolve problems and secure better control of resources.
- Lack of proper marketing support can be remedied by better marketing promotion and ensuring better prices for the producers.
- Delays in approval procedures are righted with transparency at all levels and the terms and conditions strictly adhered to.

- Too much government interference can be underplayed by pre-determining specific roles and lines of authority. The study of government should be undertaken for further remedies.
- Absence of effective evaluation by the government is overcome by submitting reports on time.
- Inadequate follow-up actions can be remedied by creating follow-up actions.

The potential area of collaboration identified by government agencies and NGOs are studied (Table No. 21).Within the broad framework depicted in earlier sections, a number of initiatives can be taken to further promote and expand Government-NGO collaboration:

Multiple and Innovative Roles of NGOs

In addition to traditional service delivery, the roles of NGOs are now more diversified particularly in response to increased vulnerability of the poor due to implementation of structural adjustment and economic reform programmes. With more cross-sector involvement for an integrated focus, NGOs are now equipped to contribute more and in new areas than before.

Capacity Building of NGOs

With diversity in functions and inter-sectoral involvement, the capacities of many NGOs, except the large and established ones with proven abilities, are stretched. Selective and appropriate capacity-building efforts of NGOs can be effective in preparing them to meet the challenges.

Government-NGO Dialogue on Development Issues

Since NGOs work mostly at the grassroots level, they have the ability to assess the impact of poverty interventions and obtain feedback from the poor. Such assessment and feedback can be effective in planning and designing strategies to address the expectations of the poor based on lessons of experiences.

NGO involvement in policy consultations and projects can be institutionalized to optimize their contribution. One way is to devise formal partnership mechanisms at various levels and/or strengthen the existing ones. Such mechanisms can be institutionalized at various levels e.g. for sectoral strategy formulation and project involvement (e.g. Government-NGO committee in each sectoral ministry dealing with poverty issues to discuss strategic plans and project interventions), project implementation (e.g. executing agency-NGO committee to discuss implementation and beneficiary feedback).

Table 21: Potential areas of collaboration identified by Government Agencies and NGOs

Organization	*Areas of Collaboration*	*Nature of Collaboration*
Government	Income generation activities	Provision of credit
	Provision of essential service packages including family planning	Establishment of service centers in rural areas

Organization	***Areas of Collaboration***	***Nature of Collaboration***
	Strengthening organizational capacities and promotion of human resource development	Skill development training
	Creation and dissemination of development information	Networking and development if information sharing mechanism
	Performance appraisal, research, evaluation of development efforts	Joint monitoring and evaluation
	Operation of micro credit	Disbursement and collection
	Health	Institutional/logistic support
NGO	Education	Institutional/logistic support
	Education (adult and children)	Institutional/logistic support, joint monitoring
	Income generating activities	Provision of credit, training, improved marketing system
	Policy making	Workshops, seminar, consultancy
	Health and family planning	Service, fund, material supply
	Legal support	Joint collaboration in human rights, services, funding & materials
	Environment	Institutional support and regulatory control by government
	Poverty alleviation	Resource allocation and consultation at all levels
	Water and sanitation	Logistic and technical support
	Women in development	Education and training
	Agriculture	Credit, technical support and joint research
	Fund raising	Joint collaboration
	Food security	Identification, plan/partnership, joint evaluation of impact
	Rural Development	Joint collaboration

Source:Study Survey, 1997

Enabling Legal and Regulatory Environment

NGOs can operate effectively and efficiently if a conductive legal and regulatory environment exists to promote rather than constrain their potential. On the part of the government, this involves e.g. devising NGO registration procedures that are simplified to induce more NGO initiatives; financial procedures to systematize rather than curb involvement; information sharing to welcome NGO input rather than to dissuade; and monitoring NGO activity to facilitate dialogue rather than control NGO activities and ensure that NGO advocacy is fruitful and not misdirected.

Alternative Financing Options for NGOs

While many NGOs are mostly dependent on grants for their activities, alternative financing options need to be explored and expanded. With greater emphasis on diversification and revenue generating mechanisms by the NGOs in recent years, provisions of credit funds and other sources to support their activities become critical.

While effective involvement of NGOs in areas where they have comparative strengths and support capacity enhancement in areas where they appear to be weak should form the basis for Government-NGO collaboration, it is important to realize certain limitations of the NGOs in promoting such collaboration. The NGO outreach is often regarded as limited and fragmented which largely excludes the extreme poor groups . Some of these limitations are the outcome of weak institutional capacity while others are conditioned by low replicability potential, e.g. good leadership, which is not often found in government agencies or even in all NGOs. In some cases, these may reflect the small size and limited availability of resources for a particular NGO or even a part of deliberate NGO strategy to achieve effectiveness by developing close relations with the beneficiaries to gain acceptability.

Moreover, such limitations may be the outcome of differing perceptions, between the government and NGOs, of success in poverty alleviation and its measurement. What is important in promoting government-NGO collaboration is to address the issues in an integrated manner to create an environment of mutual trust and understanding to harness the potentialities and advantages.

In India, successful rural development and poverty alleviation requires a coordinated strategy on three fronts: a faster growth process, human resource development; and targeted development for the poor. The traditional emphasis on macro growth strategy by the government has been increasingly supplemented by targeted development and provisions of safety nets to the poor. The NGOs, on their part, have also increasingly turned to policy advocacy in an attempt to make the growth process more poverty-focused in addition to their micro-level attempts to develop livelihood strategies for the poor. Such a reorientation of the strategies on the part of the government as well as the NGOs calls for strengthened collaboration and partnership.

From the initial phase of mutual distrust and misgivings, both the government and the NGOs have come a long way in realising their complementary role in strengthening grassroots initiatives for rural development and poverty alleviation . Although a large number of NGOs and the government are involved at the grassroots development efforts, evidences are common where little coordination among them results in duplication, overlapping and absense of concerted efforts in reaching the goals. It should, however, be pointed out that the specific goals of NGOs are unidirectional, relating to the problems of the poor whereas the Government is responsible for the entire range of issues relating to national development. Moreover, despite being non-profit organizations trying to meet socially identified needs, NGOs differ among them in terms of perceptions, ideology, concept of development and above all, funding sources.

In the past, NGOs have attributed the reasons for tension with the government to factors like differences in values and ideology, development priorities and approaches, and independence of the organizations. The government, on the other hand, maintains that managing the development efforts is the primary responsibility of the government and NGO support in carrying out the agenda must be ensured through a national monitoring mechanism. Another perceived role of the government in this respect is to coordinate the activities of a large number of NGOs with diverse ideas and strategies for a more cohesive development process.

A major element complementary of the government and NGO interventions is the need for new approaches and development initiatives and the ability of the NGOs to provide alternative models derived from a mix of local culture and tradition with new technology and management skill. To derive the maximum benefit from these diverse models, the government-NGO link must be strengthened within the broader development framework of the government.

With the gradual change of emphasis from community welfare-related activities to integrated rural development, NGOs can play an important role in complementing the government efforts through developing appropriate institutions and concomitant value system. The government emphasis on the need for extensive programme for organizing the people at the grassroots level has further expanded the scope for the partnership role of NGOs.

A range of statutory and administrative regulations exists in India for registration, prior review, project approval and utilisation of foreign funds by NGOs. The legal framework has two major dimensions: (i) Laws for incorporation and providing legal entity to NGOs; and (ii) Laws governing the relationship of NGOs with the government.

The word 'voluntary' does not suggest total absence of State control. Voluntary agencies have necessarily to operate within the framework of laws enacted by the state. They, for instance, have to comply with the Societies Registration Act, Foreign Contribution Regulation Act, Income Tax Act and the like. Their accounts are subject to audit and the government has power to investigate any foreign assistance to these agencies. The Kudal Commission enquiry into the functioning of the Gandhi Peace Foundation is a proof of the possible interface between the State and voluntary action.

The NGO Affairs Bureau can be established with the authority to register and regulate all NGOs operating with foreign funds in India. With a large number of laws, ordinances, rules and regulations applying to NGO operations, difficulties and inconsistences have emerged . The whole legal framework needs to be revamped to facilitate the promotion of a healthy NGO sector and strengthen the national context for increased government-NGO collaboration and partnership in rural development and poverty alleviation.

In view of the growing recognition of the role of NGOs in rural development and poverty alleviation, a close coordination between the government and NGOs is essential. The government and NGOs need to work in close cooperation not only to supplement each other in reaching common goals, but also to share experiences through dialogue and interaction.

The issues related to transparency and accountability of the NGOs have emerged as critical, and somewhat controversial, in developing satisfactory government-NGO collaboration. Much of the debate centering on government-NGO relations is concerned with the above issues. Currently, NGOs are accountable to the government to the donors (through evaluation and assessment), and to their own governing bodies. However, since NGO activities are financed by foreign public funds and the activities fall within the public domain, an additional accountability to the public is called for, in the interest of transparency. One mechanism is the regular publication and making available annual reports along with the annually audited financial statements to the public.

The most important aspect of accountability, however, is the transparency of NGO operations for the people at the grassroots. One way of effecting this is through participation of the beneficiaries in local and/or grassroots committees in various capacities e.g. advisory, consultative, policy planning, implementation and other follow up actions. With a carefully planned and diverse representation of the beneficiaries in all aspects of NGO activities, a high degree of transparency and accountability can be achieved.

The relation of NGOs with the local government is one of the critical factors that set the environment within which government-NGO collaboration can work at the grassroots level. This calls for effective transparency in NGO relations with local government. The practice needs to be further institutionalized and expanded to lower levels to strengthen decentralized mechanisms of government-NGO collaboration.

Within the broad framework of rural development and poverty alleviation efforts, the emphasis and focus of activities of the organizations are found to differ. The government agencies pursue their major aims and objectives of rural development;poverty alleviation; institutional development and provide assistance to relevant institutions; disaster preparedness and management; liaison with development partners; apply rules and regulations of the government; and coordinate, monitor and supervise NGO activities. The sample NGOs are observed to be involved in several areas of rural development and poverty alleviation within a comprehensive approach e.g. awareness building; human resource development; health and family planning; water supply and sanitation, promotion of environment and ecology; entrepreneurship development; research in innovative opportunities; socioeconomic development of the poor and assistance to distressed children; institutional development and assistance to social welfare institutions; and women's empowerment etc.

With the emergence of NGOs as significant partners in the development process and recognition of their contribution and emphasis on involvement, the existing government-NGO collaboration has expanded both in terms of direction and coverage. The major benefits of such collaboration at the grassroots level have been identified in terms of:

- Increased flow of benefits of development to the poor;
- Development of leadership;
- Creation of flexible functional mechanism and enhanced decision making process;
- Community participation with a sense of ownership; and

- More productive use of locally available resources.

Several forces have also been identified which constrain the satisfactory development of collaboration and partnership between the government agencies and NGOs. These include factors like multifaceted control of NGOs by, and the rigid functional mechanism of, the government, inadequate communication resulting in lack of coordination and understanding; low transparency and accountability; and unnecessary interference. The suggested measures indicate that interventions should involve a number of areas:

- Existing rules, regulations and systems of the government;
- Cooperation and coordination;
- Project management; and
- People's participation.

The changing of rules, regulations and systems, however, does not necessarily ensure the development of a conducive environment for enhanced collaboration. What is necessary is to evolve a system, where both the parties respect certain boundary conditions and abide them. The cooperation and coordination are necessary at all levels by the respective stakeholders. This may not be comprehensive but suggests a wide range of areas and activities where effective collaboration can flourish.

Experiences of NGO-Government Collaboration in Marathwada

The NGOs visited and observed in this day have a evemplary partnership with the government agencies. Government agencies and the semi-government agencies support these agencies, with technical assistance, funds, guidance and also involve the agencies in various government programs for rural development, viz., DWCRA, water supply, watershed, DPAP, soil and water conservation, MCED, social forestry etc. There are some agencies like IHMP, which provide training and guidance to various government departments of State and the country. This partnership between government and voluntary agencies in Marahtwada is cited below.

Manavlok has initiated an awareness program on water, sanitation and environmental care in 46 villages under the guidance of Maharashtra Rural Water Supply. The agency also received funds from PAD Maharashtra, semi-government agency like CAPART Delhi and Zilla Parishad Beed provided support for construction of bio-gas plants. It also received Zilla Parishad Beed Award for biogas units.

Grasp is an active partner of the governemnt. Its emphasis is on collaboration with government department for supplementing rural poverty alleviation programs, focusing on the strengthening, planning, co-ordination and monitoring of local governance by expanding the involvement of the community in implementing various activities viz., DPAP, Jalsandharan, Adarsh Gaon etc. Grasp is a member of the district Advisory Committee for DPAP and Model Village Scheme of government. It interacts continuously with the government officials to enable changes in government monitoring and sanctioning process.

Janarth, the government recognised various educational programs of this agency viz., balwadis, study centres, science laboratories in project areas, akshar yatra etc and extends help as required. It has extra-curricular activities in collaboration with government secondary schools. The agency carried out tree plantation program with the help of the Social Forestry Department of the government. In 1996 it was undertaken by government DPAP program for watershed development.

Nirman has strengthened its health activities by networking with various hospitals in which government hospital of Aurangabad has played a significant role. The agency in association of NABARD and IGWDP works for its natural resource management. It has also conducted technical and social surveys for Soil and Water conservation projects of governemnt of Maharashtra and IGWDP.

Abhinav Vikas Sanstha (AVS) this agency was undertaken by IGWDP in 1999 for watershed developemnt program in Varkhedi. Indo-German Social Service Society and NABARD supported the agency.

Institute of Health Management Pachod (IHMP) its Pune centre is set-up to develop low cost audio-visual training material for the government sectors and NGOs. It develops material in reproduction and child health, water, sanitation, nutrition etc. It has trained over 2500 government and NGO functionaries in 1986 through practical oriented courses. It developed information, education and communication (IEC) model for the decentralised communication strategy for Government of India (GOI). This provided the basis for the national policy for the decentralised IEC, formulated by GOI in January 1999. It also undertook Training Need Assesment and training of 29 DMOs and 11 trainees from the state institutes and IEC bureau for the State of Maharashtra.

Dilasa is engaged in implementing DWCRA, a sub scheme of IRDP in Aurangabad district. The National Instituite of Rural Development appreciated its performance in DWCRA group formation. Therefore the government entrusted the responsibility of IRDP for self guarding the neglected section of women to this agency. It assists the villages in income generating activities on various government development schemes. It plays a catalytic role in socio-economic development of villages by coordinating various government agencies and the villages together. It also implements watershed development programs under DPAP and IGWDP.

References

The first Five Year Plan, 1951.p. 150

Sharing on Earth: Extract from a report of FAO-FFHC/An Asian Consultation on responding to the challenge of rural poverty in South Asia: The Indian journal of Public Administration, Quarterly journal of the Indian Institute of Public Administration July-Sept 1987,Vol XXXIII, No.3. pp.784-786.

Swapan Garain: Government- NGO Interface in India – An overview, Indian journal of Social Work, Vol. No. 3, (July 1994).

Fernandes A.P.1987: NGO's in South Asia: 'People's Participation and Parternship,' World Development Vol. 15. pp.39-41.

Government of India 1957: Report of the team for the study of Community Projects and National Extension Service, Vol. 1, New Delhi 107. (Balwantrai Mehta Committee).

Sanjit (Bunker) Roy: Voluntary Agencies in Development: The Indian journal of Public Administration, Quarterly journal of the Indian Institute of Public Administration, Special Number on Voluntary Organization and Development, their role and function. p.461.

Recognition of Voluntary Sector in Five year Plans and in the approach Paper to the tenth Plan; Report of the Steering Committee on voluntary Sector for the Tenth Five Year Plan (2002-2007). Planning commission Government of India (Jan 2002), Source –Internet.

Ibid.

Ibid.

H. Srikanth: State, NGO's and Urban Community, Social Action, A quarterly journal of Social Trends, A social Action Trust Publication, Jan- March 1996, Vol.46. No 1, pp. 42-45.

ASIAN NGO coalition for Agrarian Reform and Rural Development 1984, Status Papers on NGO involvement: A Perspective of ten Countries in Asia, Manila; ANGOC.

Economic and social Commission for Asia and Pacific 1989: The Socio-Economic Impact of Rural development Program on Low income Groups, New York, ESCAP.

Garlio E.D.1987: Indigenous NGOs as Strategic Institution's Managing the Small Relationship and Government and Resource Agencies, world Development, Vol.15.pp.113-120.

Sanjit (Bunker) Roy: op cit. p. 459.

Ibid. 460.

Rajalaxmi. T.R (1993) 73rd Amendment: A Window for Alice? Mainstream Sept.18. pp. 18-21.

Ajay Dubey: Strengthening Rural Local- Self Government (Panchayat Raj) Institutions, ' Strategies for NGO's Journal Rural Reconstruction 29(1) 1996.

Ibid.

Government-NGO Cooperation for Poverty Alleviation: Centre on Integrated Rural Development for Asia and Pacific, (CIRDAP), Table 4.2 Government – NGO Collaboration: Problems and Suggested Measures, Source study Survey, 1997,p.56.

Government-NGO Cooperation for Poverty Alleviation: Centre on Integrated Rural Development fpr Asia and Pacific, (CIRDAP), pp.46-48.

Ibid.p.49.

Shriram Maheshwari: Voluntary Action in Rural Development in India, The Indian journal of Public Administration, Quarterly journal of the Indian Institute of Public Administration July-Sept 1987,Vol XXXIII, No.3.p.560.

Government-NGO Cooperation for Poverty Alleviation: Centre on Integrated Rural Development for Asia and Pacific.pp.54-55.

CHAPTER - 8

Findings and Conclusions

Voluntarism in India is as old as the emergence of organised society itself. It originated as pure philanthropy of charity and this motivation sustained the voluntary efforts all through history in the ancient and medieval period. The voluntary efforts in the process of welfare and development have undergone evolutionary changes with changing emphasis on various experimental development programmes in India. The history of voluntary action is an integral part of the study of evolution and changes in the Indian society. The five sources of voluntarism are religion, government, business philanthropy and mutual aid. Mutual aid and philanthropy are the two main sources from which voluntary associations developed. The missionary zeal of religious organisations, the commitment of government organisations to the public interest, the profit making urge in business, the altruism of the social superiors and the motive of self-help among fellowmen , all reflect in voluntarism. Tracing the development of voluntarism through eras it is studied that in ancient and medieval times, voluntarism had its conception in Dharma and the ambit of Sasana (directives of king). They promoted moral, aesthetic, material and spiritual progress of the whole community.

In India before independence development was instituting and extending facilities for agriculture, health and medicine, education and allied human development. In Muslim era the role of voluntarism in India was not very much affected. During British rule traditional sources of voluntarism in development received a boost. Missionary hospitals and schools were established and the traditional pathshalas and madrassas were pushed to the background. In this era voluntarism played a significant role in educational, health and medical and social welfare development in India during the later half of the nineteenth and the first half of 20th century.

Voluntarism secured a fresh lease in the National Swadeshi Movement. Gandhiji based it on the philosophy of spiritualism of the soul force or love force, which to him mark the Indian culture from the western. Voluntarism was the core of reconstruction of economic and political organisations. After independence voluntarism receded into a negligible role in development because the assistance came from government through its Five Year Plan outlays.

In Maharashtra rise of voluntary service was brought about by many political activists, philanthropist and social reformers. They all worked for common man, rural people, women, education and upliftment of untouchables. These social reformers were Mahatma Phule, Justice Ranade, Gopal Ganesh Agarkar, Bal Gangadhar Tilak, Pandita Ramabai, Dr. Babasaheb Ambedkar, Vithal Ramji Shinde, Sayaji Rao Gaikwad, Shahu Maharaj, Keshav Karve, Karmaveer Bhaurao Patil, Punjabrao Deshmukh and many more.

In the history of voluntary agencies in Marathwada the contribution made by the social reformers like Swami Ramanand Teerth, Govindbhai Shroff, Shri Paranjpayee, Borikar, Pedkar, Charthankar, Chapalgaonkar, J Desai, Capt. Joshi are pioneering.

An immense contribution was made by the Christian missionaries in the field of relief, health, agriculture, education, watershed, establishing War-on-Want. They devised the pioneering model for boring apparatus and devised a new technique for boring wells. Marathwada Sheti Sahayya Mandal (MSSM) the voluntary agency studied in this research is formed from the War-on-Want and American mission and local philanthropists.

Rural development is a process which includes the development of socio-economic conditions of the people living in the rural areas, and ensures their participation in the process of development for complete utilisation of physical and human resources for better living conditions. It extends the benefits of development to the weaker and poorer sections of rural society. It also enhances both the capacity and capability of administrative and socio-economic development agencies and agricultural marketing units working in the rural areas.

The efforts made for rural development generally failed to bridge the gap between the rich and the poor. Moreover, the benefits of the schemes reached not all the villages. This was the problem of rural development. The government agencies and political parties failed to bring in people-centered rural development.

Due to the failure and weakness of government policies for rural development the NGOs come to act as an intermediary development agencies between the State and the target groups by acting as their guide, philosopher and friend.

NGOs are more target oriented, flexible in their methods and practices, and adopt innovative and participatory approaches. They are more focussed and effective in developmental work and enjoy independence. Advocating and lobbying are better done by NGOs because they can influence the perception. Voluntary agencies assure paramount importance in the rural development. The various rural development schemes, poverty alleviation programs and minimum needs programs are also some projects of the NGOs. The rural planning and its execution have been given to the people by the way of PRI's. In this decentralized planning, the role of NGOs has become more important than before.

NGOs organised beneficiaries into co- operatives at the village level so that they could secure reasonable prices for their products. They help at the grass root level to motivate beneficiaries to change their occupations from traditional activities to Industry Service Business (ISB) Sector. NGOs play a vital role in the proper implementation of the JRY by creating the awareness among the rural people. The NGOs promote various types of innovative projects at village, block and district levels. The NGOs help the local youth, farmers and women institutions by way of providing technical/professional help in carrying out the projects on behalf of the village community. It identifies dedicated youths, kisans, mazdoors who have leadership qualities, this should be inducted in the respective organisation. The NGOs encourage the people to generate more non-conventional source of energy and use the same effectively in the production function. They co-ordinate various schemes of different departments at the village level acting as a catalyst. NGO's are involved in the village areas for technological development like selecting relevant technology from different alternatives; making it adaptable; providing science and technology and support for its successful commercialization.

There are various programs where the government involves NGOs for implementation and management viz.,Development of Women and Children in Rural Areas (DWCRA),Training of Rural Youth for Self Employment, Drought Prone Areas Programme (DPAP), Desert Development Programme, Disaster Relief, Drought Relief, Flood Relief, Rural Women and Child health Programmes, Family Planning Programmes etc.

The concept of NGO, the importance given to the NGOs and their contribution/ role expectation of the government in development efforts have undergone a significant change over different Five-Year Plans. Both in partnership can bring about complete rural transformation.

Maharashtra is the third largest State in India. Half of Maharashtra is drought prone. This results in degradation of agricultural land, loss of productivity and scarcity of food, fuel, fodder and water. Recurrent droughts with their negative impact on communities and socio-economic conditions has led to a search for sustainable solutions. The government and voluntary agencies have tried for many years to investigate the effects of droughts by adopting sectoral approach which was ineffective. It also proved that watershed development can be the only comprehensive and sustainable strategy for rural development.

Marathwada has been undertaken in the study because this region is relatively backward in the state. There are varies factors causing this backwardness., viz socio-economic, political, educational, technological, agricultural etc. One of the major component of rural development is agriculture. As this region lies in the drought belt of Maharashtra, water and its allied developments are very important. It is found that after the liberation of Marathwada the development of the region was negligible. The government due to many factors avoidable and unavoidable has been unable to make much inroads into the development process of this area. The emergence of few NGOs in the area are found to be making significant progress and impact in their localised areas of operation. The need for many more, larger and effective NGOs working in different sectors of development is very necessary. An attempt has been made to review a few NGOs in this region.

NGOs in Marathawada

Manavlok operates in Beed district, Ambejogai Tehsil in Marathwada. The objectives of this agency are to improve the socio-economic status of the rural people of the region, to empower women economically, socially and politically,to improve health conditions in the region,to improve the standard of formal and non-formal education and increase the academic standing of student to build democratically working transparent organisations of rural poor, women and youth to solve their own problems locally.

Some of the important activities carried out by the agency are; soil and water conservation, mobile health clinic, community wells for marginal farmers, legal support, skill training for women, land-less and youth, forming Krushak Panchayats for farmers and Bhumikanya mandals for women.

Nirmaan operates in Aurangabad and Jalna districts of Marathwada. Objectives of this agency are; health for all, essential education and literacy, land capability based agricultural development conservation and management of natural resource, Livelihood development, gender and equity, resource support, technology development. Some succesful activities carried out by this agency for urban and rural areas are health, natural resource management like watershed and resource reclamation agricultural support and training for dry land farming.

Jannarth is an agency working in Gangapur Taluka in Aurangabad district. Objectives of this agency are; women empowerment, preventive and rehabilitative medical work, economic development through income generation programs and marketing facilities, to emphasise the preventive and promotive aspects of health care, to promote education, to enrich the environment by natural resource management, to empower marginal farmers by maintaining agricultural service centre.

The major focus of health activities are medical rehabilitation and prosthetic distribution. Ayurvedic practice and awareness programs were carried out and ayurvedic laboratory to test the active constituents of herbs was set up. The other important activities are warehousing and marketing assistance, natural resource management, watershed management and plant nurseries, urban activities, Ayurvedic clinic, balwadis, sports groups and working with rag pickers.

Abhinav Vikas Sanstha works in Soegaon, Aurangabad, which is the most backward block of Aurangabad district. Ojectives of this agency are; Awareness and mobilization, people's initiatives, natural resource management, gender sensitization, skill and capacity building, credit linkages and agriculture promotion. The activities carried out are women's development, forming self-help groups, watersheds, youth activities.

Grasp is an agency working in 7 districts of Marathwada. Objectives of grasp are; to develop strategic focus in identifying the technical and organisational management needs of grassroots level organisations, to strengthen the management and utilisation of the natural resources and to regenerate the degraded lands to fulfil the demands of the local community, to build capacities of the grassroots level organisations and CBOs and promote new grassroots organisations to work for ERM, to develop areas and regions plan to implement and replicate the WDP/ ERM programmes, to develop the strategy for research, network building and advocating the policy for broad basing the concept of community empowerment using ERM as a tool. Activities carried out by Grasp are community action projects, forest protection, community banking system, anti-alcoholism, social fencing and joint forest management.

Institute of Health Management Pachod (IHMP) works in 72 villages in Paithan Taluka, Aurangabad district. Objectives of IHMP are; the holistic development of the individual, family and community, upliftment of marginalised groups,the health and development of women, children and most disadvantaged, mobilising communities toward self-reliance and sustainability and to organise children and adolescents, through devlopment of health, water and sanitation programs. Some of the important activities carried out by IHMP are primary health care, Balvikas, child centred development, disaster management and policy advocacy.

Dilasa works in villages of Aurangabad district. The prime objectives of Dilasa janvikas Pratishthan is women empowerment and self-reliant, self sustainable development of villages. The organisation works for these objectives through;Development of Women and Children in rural Areas DWCRA focussing on the women living below poverty line, ecological concerns and conservation and management of natural resources and forestry, land capability based agricultural development, school sanitation, resource support and technology development and gender equity and education of women. Activities carried out by this agency are implementation and monitoring of DWCRA, rain water harvesting, watershed, social forestry, vermi-culture etc.

MSSM is a NGO in Aurangabad which implemented a watershed project in Adgoan village. It is found that the village has been totally changed after the watershed project implementation. People have acquired a attitude for development of their own self and the village. Today there are 25 self-help groups in which there are 23 groups of women and 2 men. Today also a watershed development committee, which was started in 1986 by MSSM is working effectively towards the management and repairs of the watershed. The annual income of every farmer has increased nearly from 4 - 5 lakhs. It is now difficult to find labourers in the village because, those who did not have land before the watershed have got their own land now. Dairy is one of the distinguishing feature, which was formed in 1982 with due guidance and leadership of Vijay Anna Borade. Now there are 3 more dairies in the village, one being run by the women. All these dairies constitute 1700 litres of milk daily. The days income varies from 2 to 2.5 lakhs. Dairy has brought joint venture business for the farmers in village, people got labour by bringing fodder, increased livestock and fertile land due to cow-dung fertilizer.

The aspect of people's participation in this project is the most outstanding feature. People were involved in all stages of project management including planning, formulation, appraisal, implementation, monitoring and evaluation.

Undoubtly people's involvement in the implementation of various activities is central to the ultimate success of the project. It has been observed in the study that out of the many selected NGOs the most successful agencies have been those who have involved the local people. The success of an agency depends on the people's desire to change and the total involvement in the working, maintenance and supervision of the project.

Integrated watershed development requires co-ordinated efforts of government agencies, voluntary agencies and village community towards achievement of the object. Hence it is imperative that a strong leader or agency should involve to take the responsibility of the co-ordination. This aspect is the stricking feature in the Adgaon project. The credit of the fieldwork goes to Shri Vijayanna Borade and the planning success to Shri J.M. Gandhi. These motivated leaders who had vast experience of higher technology brought superiority to the project implementation. Their exposure to urban areas and atmosphere lent quality to the project. They have qualities of dedication and sacrifice which is the most essential element for the success of any project. "MSSM has a very dynamic leadership of Shri Vijayanna Borade and if the government is supported by such leader in their project government can change the whole scenario of the rural sector." (Interview with the collector of Aurangabad).

One of the laudable endeavours of the project is the change in the women's lives. Though the women were not involved in the planning and policy making part of the project, they were actively involved in the implementation and management of the project. Due to the sincere and hard efforts of the women Adgaon has seen economic and social enhancements. The women training programmes for skills and technologies has increased their self-confidence and enabled them to undertake work in all the activities of the development programme including the male dominated ones. The study observed that the women were over burdened with the household activities and the field work however were able to cope with the extra burden. It is suggested that to facilitate more of the women involvement and for the change in attitudes to inculcate leadership in women, skill development training and accreditation of existing technologies specially for women is essential. The leadership qualities should be identified and nurtured. These women should also be involved massively in policy making of varous projects.

India has witnessed a positive transformation in women's empowerment. In 1971, the government appointed a commission on the status of women in India, so that the women were looked upon as the actors of development. But in the fourth and fifth plan it was clear that there was no mention for women. In 1975, Lok Sabha passed a resolution to initiate comprehensive program for women. It was the Sixth Plan, which can be taken as a land mark for the women. A separate Ministry for Women and Children was created in 1984. The various programmes for women development were IRDP, DWCRA, TRYSEM, 41 items under EGS which included NREP, RLEGP, JRY. Etc.

In the process of empowerment of rural women NGOs play a very significant role. There are many agencies which take up the challenge of empowering women. Adgaon project of MSSM has striking features regarding involvement of women in development activities and process. Their contribution to the project provided wage employment to the women. There is a lot of work for women on their own farms due to the increase in crop cultivation and pasture development activities enabled them to have supplementary income. Today there are women involved in the watershed development committees. They are actually involved in the management process, there are 3 mahila mandals in the village, 22 self-help groups and one dairy run by women (Saraswati Dairy). Then the Sarpanch of Adgaon Gram Panchayat was a woman.

Dilasa has a range of activities for women development. It has a multipurpose women centre. The objectives of this centre is to give legal aid, family counselling, income generating activities and rural enterpreneurship for women. This organisation has formed 22 womens co-operative dairies.

Abhinav Vikas Sanstha has been tackling the problem of womens low literacy rate and exploitation of women. The organisation developed self-help groups among women were many members made savings of 4 lacs of rupees.

Nirman started self-help groups and mahila mandals to strengthen the women. It also organises in house training, special training for leadership development, help, education, gender development and other trainings to empower the women of the area.

The special women oriented project of Manavlok is Manaswini. The rural women especially the poor are organised to form Bhumikanyamandals' in which the members come together to discuss, solve their problems and join Krushak Panchayats in development programmes. It also lends legal advice and support for destitute and deserted women.

Janarth has awareness programme for women to build confidence in the women by increasing their knowledge base. The various programmes undertaken are formation of mahila mandals, incentive schemes for participation in mandal activities are also included like, seed gathering, medicinal, nursery raising, collecting of medicinal raw material. This organisation also conducts workshops for women and girls on their legal rights.

Grasp has promoted 21 self-help groups. It also works for gender sensitization focussing on empowerment of women and enhancing their participation in the developmental process. The organisation ensures womens participation as stake-holders in the management of watershed development. They also promoted saving groups for women which followed into savings and credit and then to enterpreneurial activities and community banks. This organisation understood the importance of women participation in development and they empowered women by increasing their participation in decision-making levels and take up income generating activities. They provided them with infrastructural support and interacting with women through one to one bases and through the community.

The rural poor, who form a majority in the developing countries should be empowered so as to regain ownership and control of resources, like land, water and forests. There is an urgent need for organizations which can work towards this goal. NGOs could play this role of an alternative model of development and work on a common program with the State support and collaboration. NGOs and government relationship is supportive as well as controversial in many areas. NGOs in India are operating in a situation of distrust and suspicion with the government machinery especially at the local level.

Role of NGOs in development got a boost in the seventh five year plan when it was declared that efforts were being made to involve NGOs in various state development program and to supplement the government efforts. There are four forms of government NGO co-operation which are conceived. They are source of information, consultants, executing of cooperative agencies and co-financiers. The NGO support the government sponsored poverty alleviation programs in ; Financial interventions, grassroot organization support services, in social and women's development, in resource conservation, natural resource management and in disaster management. The qualities of a good NGO- GO merger should be able to supplement government efforts to be the eyes and ears of the people, to activate the system and make it respond, to disseminate information, to make communities as self reliant as possible, to train cadre of grassroot level workers, to mobilize financial resources, to mobilize and organize the poor. The NGO-GO collaboration has three major types of arrangements, the sub contract, joint implementation and government as financier of NGO projects. The commonest collaboration is the sub-contracting arrangements where government agencies enter into contract with NGOs.

The problem and suggested measures in the NGO-GO collaboration is lucidly explained. The measures suggested for effective NGO-GO collaboration for government are to create a permanent platform at national and local level for mutual interaction, to form conducive policies and laws and to launch more of joint projects, to form environment to facilitate healthy and effective growth of NGO activities and to ensure the transparency for NGOs, to set up proper and open communication systems, to make written contracts with partners, should have frequent meetings and consultations with each others. NGOs should directly contact the concerned authority in the bureaucracy to avoid corruption and red-tapism, NGOs should provide technical and developmental skills for the emloyees, they should have transparency at all levels. NGOs should study government programmes, together discuss and solve the problems and should submit required reports to the government on time.

Following are the suggestions for the effective working of the voluntary agencies;

Any development programme in rural areas will be successful only when the local people are properly informed about its benefit. Involvement of the local people in the management of the project is necessary. Therefore to initiate any project NGOs should mobilise local people and enlist their support to contribute in planning, implementation and evaluation of the project. This process will help rural people not only to become self-dependent but also affirm their unflinching commitment to the project by the way of active participation.

The main beneficiaries of watershed projects will be the rural women. They can assist in mobilising support for the projects, raise the initial capital and in many cases contribute most of the labour. Rural womens involvement in the watershed project programme will have a long lasting impact. So the NGOs should make an organised effort for emancipating rural women, which can make them mentally strong and enable them to take up social responsibilities. They must reorient the women's training programmes particularly which are of non-conventional type. Rather than the programs and supply schemes of sewing machines, tailoring, papad making, toy-making, etc the programmes should be evolved in a way to provide an armor for women to take up new challenges. Hence NGOs must take up non-formal education to create awareness and to built up a new cadre of rural women for undertaking voluntary work.

Training local people to operate and maintain a project, is an area which requires active attention of NGOs. To train local people in planning, project management, project evaluation, record maintenance must be the priority object of every NGO working in the rural sector to make the people self- reliant. Any rural development should lead the masses to self-reliance. Educating the rural masses on their lawful rights and helping them in acquiring those should be an essential object of the awareness programmes of the NGOs.

NGOs should also set up youth forums for self help purposes and for abolition of the detestable social systems, practices and addictions. This forum can also create conditions for rehabilitation of addicts.

The NGOs must be more accountable to the funds received from various funding agencies, especially the State agency. They must keep in mediating for rural masses in the banking procedures. Lastly the NGOs must ensure economic stability and self-reliance before they withdraw the area of operation.

The study observed that the NGOs have made an effective and sustainable development in many areas of Marathwada. The philanthropic, professional attitudes of the NGOs and their active collaboration with the State has brought about radical transformation in the rural region of Marathwada. The dedication of the NGOs to the drought areas, made them economically, agriculturally, politically and socially developed villages. Marathwada needs more of such effective NGOs for its development in various sectors. With the effective collaboration of the State and NGOs and with a paradigm shift of the thinking, planning, and execution of the projects, Marathawada will no longer remain the most backward region of Maharashtra. With highest participation of people, government agencies, leaders and donors, NGO's would become the backbone of Marathwada's holistic rural development and transformation.

Bibliography

1. Amartya Sen in The Hindu] Delhi 6 Nov. 1995, Interviewed by Rammanohar Reddy.
2. Amartya Sen : In The Hindu (Delhi), 6 November 1995. Interviewed by Rammanohar Reddy.
3. Ajay Dubey : Strengthening Rural Local-Self Government Panchayati Raj Institutions, Strategies for NGO's' Journal of Rural Reconstruction 29 (1) 1996.
4. Akshay Sood : ASHWATTHA] The bodhi tree that symbolises holistic knowledge and the universal man Quarterly Journal of YASHADA, Vol. 3 No. 1 - 3, January - September 2000.
5. Amrita Basu, Two faces of Protest : Contrasting Modes of Women's Activism in India] Berkley University of California Press, 1992 , p. 16.
6. Alka Srivastava : Non-Governmental Organizations and Rural Development : Social Action; A Quarterly Review of Social Trends; A Social Action Trust Publication, Vol 49. January-March 1999. pp. 29-30.
7. Alka Srivastava : Non-Governmental Organisations and Rural Development Social Action, Vol 49. Jan-March 1999, pp. 27-44.
8. Anna Saheb Garud and B.B. Sawant, History of Social Reforms in Maharashtra from 1818-(1950), Kailash Publications, 1995. pp. 98-102.
9. A. Major, Pilot Project in India, University of California Press, 1958, p. 37.
10. A. D. Jedlica, Organisation for Rural Development: Risk Taking and Appropriate Technology, New York, Praeger, 1977. Also see C. Bryant and L. G. White, Managing Development in the Third World, Westview, Boulder Co., 1982.
11. A.F.C. Bourdillon (ed) Voluntary Social Services : Their Place in Modern State. London, Methuen, 1945. p.2.
12. A.F.C. Bourdillon, op.cit, p. 7.
13. A.S. Altekar, State and Government in Ancient India, Delhi, Motilal Banarasidas, 3rd Edn 1958. (reprint 1972) p. 384.

14. B.B. Mishra : District Administration and Rural Development, Delhi, Oxford University Press, 1983, p. 6.
15. B. Rambhai : The Silent Revolution, Delhi, Jiwan Prakashan, 1959, p. 10.
16. Baumgartner, Ruedi (1989); Participative and Integrated Development of Watersheds : Reflections on Experiences from an Indo-Swiss Collaboration, Swiss Development Corporation, Banglore.
17. Bina Roy : The Status and Role of Women in Our Changing Society - Social Development Essay's in Honour of Smt. Durgabai Deshmukh, edited by B.N. Ganguli, Sterling Publication Pvt. Ltd., pp. 58-59.
18. Batliwala S. (19930 : Empowerment of Women in South Asia : Concepts and Practices Sponsored by the Asian South Pacific Bureau of Adult Education and FAO's Freedom for Hunger Campaign, Action for Development, Banglore.
19. Boutros Boutros Ghali : Secretary General of the UN, in his welcome speech on the Fourth Congress on women in Beijing.
20. Baidyanath Misra - Poverty, Unemployment and Rural Development. Ashish Publishing House. Delhi 1991.
21. Balkrishna, Lalitha and Mishra Meenu : 1998, Role of NGO's in Sustainable Rural Development : IREDA News, Vol 3, July-Sept 1998, p. 47.
22. Batliwala S] 1993 : Empowerment of Women in South Asia : Concepts and Practices sponsored by the Asian South Pacific Bureau of Adult Education and FAO's Freedom for Hunger Campaign, Action for Development, Bangalore.
23. Bagchee Sandeep., 1987, " Poverty, Alleviation Programmes in Seventh Plan : An Appraisal", EPW vol. 22 (4) pp. 139-149
24. Bharat Jhunjhunwala, Voluntary work as Contervailing Power', Economic and Political Weekly 21, 1986, p 599.
25. Bhasin K. 1992. Education for Women's Empowerment : Some Reflections. Adult Educational Development, March, No. 38.
26. Bhasin] 1992 in K.C. Vashista's and Sashi Malik's : Some Strategic Effort Towards the Empowerment of Women, Conceptual Framework of Women's Empowerment; University News, Vol 4, No. 5, Feb. 4-10, 2002, p. 12-16.
27. Bhatt R.V. " Maternal Mortality in India, Project Report : 1992-94
28. Bhatt R.V. 1995 - Maternal Mortality in India, Project Report 1992-1994 - Personal Communication.
29. Bhatt R.V : Maternal Mortality in India. Project Report 1992-94, 1996.
30. Bina Agarwal 1994, Gender and Command over property : A critical gap in Economic Analysis and Policy in South Asia, in World Development, vol. 22, n.10, p.1455. 1994
31. Bedi, Narinder, 1999, Development of Power, In D. Rajasekhar] (ed) Decentralised Government and NGO's : Issues, Strategies and Way Forward, Delhi, Concept.

32. Bruce F. Johnson and Peter Kilby, Agricultural Strategies : Rural-Urban Interactions and the Expansion of Income Opportunity, Paris, OECN, 1973, p. 15.

33. Bedi, Narinder, 1999, Development of Power, in D. Rajasekhar] ed., , Decentralised Government and NGOs: Issues, Strategies and Ways Forward, Delhi, Concept.

34. Bhat M. K., 1999, Sustainability and Local Governance, In D. Rajasekhar] ed., , 1999.

35. Boutros Boutros Ghali - Secretary General of the UN, in his welcoming speech on the fourth World Conference on Women, Beijing. 1995

36. Benajamin U : Bagaden and Francis, F. Korten, Developing Irrigators Organisation : A Learning Process Approach' in Cernea] (ed) Putting People First. p. 85.

37. B.L Agarwal - Issues in Rural Employment. Chapter in Rural Development. Anmol Publications, Delhi.

38. B. Sahoo. Approach to Rural Development :Ashish Publishing House, Delhi.

39. Baumgartner, Ruedi (1989) : Participative and Integrated Development of Watersheds: Reflections on Experiences from an Indo-Swiss Collaboration, Swiss Development Corporation, Bangalore.

40. Chambers, Robert (1994): Participatory Rural Appraisal] PRA : Challenges, Potentials and Paradigm, World Development.

41. Constance Smith and Ann Freedman, op.cit, p. VIII.

42. C. Francis, Rural Development, People's Participation and the Role of NGO's. Journal of Rural Development, Vol 12 (2) 1993, NIRD Hyderabad] India . p. 205.

43. C.S. Nagpal and A.C. Mittal : Rural Development Perspectives in modern Economics - 13. Anmol Publications, New Delhi

44. C.N. Ray. " Politics of Rural Development - Rawat Publication 1994. Delhi

45. Cohen, C. Demoracy] New York : Free Press 1971. p. 7.

46. David L. Sills op.cit, pp.362-363.

47. Dr. A.M. Kathare and Dr. N.N. Nagrale, History of Marathwada, Beginning to (1960) : Kalpana Publications : 1999] marathi edn p. 56.

48. Dr.Mrs. Sudhatai Kaldate; Social Change in Marathwada; Development of Marathwada - A Perspective; Publication, Swami Ramanand Teerth Research Institute, Aurangabad, 1999, p. IX.

49. Dr. Kurulkar R.P.; Economic Development of Marathwada; Development of Marathwada - A Perspective; Publication, Swami Ramanand Teerth, 1999, p. 38.

50. D.D. Gow and J. Vansant, Beyond the Rhetoric of Rural Development Participation : How can it be done ? World Development, Vol II, 1983, pp. 51-54.

51. Dr. Mrs. Sudha Tai Kaldate; Social Change in Marathwada : Development of Marathwada; A Perspective; Publication, Swami Ramanand Teerth Research Institute, 1999. p. IX.

52. Durgesh Nandini, Rural Development Administration, Ravat Publications, Delhi 1992.

53. Douglas Houghton - Priorities in Voluntary organisation, London 1967.

54. D.L. Sheth and Harsh Sethi, The NGO Sector in India : Historical Context and Current Discourse,' Voluntas 2, 1991. p. 51.

55. Dr. P.M. Bora, Peoples Participation in Food Administration In India : A Study of an Indian State 1982 A Janta Publication. pp. 201-202.

56. Dr. S. Panda - Rural Development : Some Fundamental Issues and search for Solutions. 1991

57. Dr. R.P. Kurulkar, Economic Development of Marathwada : Development of Marathwada; A Perspective; Publication, Swami Ramanand Teerth Research Institute, 1999. p. 38.

58. Davis George : Dynamics of Power : The Gandhian Prespective, Frank Bros. & Co. New Delhi, 2000.

59. Dixon, John M (1977): Farmers' Participation in the sustainable Development of Natural Resources in Rainfed Areas in Prem N. Sharma] (ed), Participatory Processes of Integrated Watershed Management, PWMTA - FARM Field Document No.7, Katmandu, Nepal, Netherlands/UNDP/FAO.

60. Desai A.R. (ed) 1971. Essay on Modernisation of underdeveloped Societies, vol I & II - Popular, Bombay.

61. Dholakia R.H. and S. Iyengar, 1998, Planning for Rural Development : Issues and Case Studies, Bombay, Himalaya Publication House.

62. Douglas Houghton - Priorities in Voluntary organisation, London 1967.

63. Durgesh Nandini : Rural Development Administration (ed), Conceptual Framework of the Study, Rawat Publications, Jaipur, 1992, pp. 10-11.

64. D. Rajasekhar, Non-Governmental Organisation] NGO's in India. Opportunities and Challenges; Journal of Rural Development, Vol 19 (2), NIRD Hyderabad, pp. 249-250

65. Eldersveld, Jagannathan and Barnabas, The Citizen and the Administration in a Developing Democracy, Illinois Scot Foresman Co, 1968, p. 4.

66. Edwards, Michael and David Hulme] (Eds.), 1995, Non-Governmental Organisations - Performance and Accountability: Beyond the Magic Bullet, London, Earthscan.

67. Farington, John, et.al., (Eds.), 1993, The Reluctant Partners? Non-Governmental Organisations, the State and Sustainable Agriculture, London, Routledge.

68. Fernandez, Aloysius P., 1996, The MYRADA Experiences: Working with the Government in Multilateral and Bilateral Projects, Bangalore, MYRADA.

69. Fernandes A.P. 1987 : NGO's in Aouth Asia : People, Participation and Partnership', World Development Vol 15, pp. 39-41.
70. Fernandez, A. P.] 1994 : The Myrada Experience: The Intervention for a Voluntary Agency in the Emergence and Growth of Peoples' Institutions for Sustained and Equitable Management of Micro-Watersheds, Myrada, Bangalore.
71. Fan, Shenggen, and Peter Hazell] 1999 : Are Returns to Public Investment Lower in Less Favoured Rural Areas ? An Empirical Analysis of India, EPTD Discussion Paper No. 43, IFPRI, Washington.
72. Fan, Shenggen, and Peter Hazell] 1999 : Are Returns to Public Investment Lower in Less Favoured Rural Areas? An Empirical Analysis of India, EPTD Discussion Papers Number 43, IFPRI, Washington.
73. F. Fama and M. Jensen, Seperation of Ownership and Control', Journal of Law and Economics 26. 1994. p. 301.
74. Fred W. Riggs, The Idea of Development Administration,' in Edward W. Weidner, Development Administration in Asia, Durham, N.C. Duke University Press, 1970, p. 25.
75. Fernandiz, Aloysius P. 1996 : The MYRADA Experiences : Working with the Government in Multilateral and Bilateral Proejcts, Bangalore-MYRADA.
76. Farrington, John et al] (ed) 1993, the Reluctant Partners ? Non-Governmental Organisations, the State and the Sustainable, Agriculture, London Reutledge.
77. Gro Halem Brundtland, Prime Minister of Norway at the Fourth World Congress on Women 1995 - Beijing.
78. Garilao E.D. 1987 : Indigenous NGO's as Strategic Institutions' Managing the Small Relationship and Government and Resource Agencies, World Development, Vol 15, pp. 113-120.
79. G. Honadle and R. Klaus, (Eds.) International Development Administration - Implementation Analysis for Development Projects, New York, Praeger, 1978.
80. G.D.H. Cole, Voluntary Social Services',] Edited by A.F.C. Bourdillon London. Methuen. 1945. pp.11-12.
81. H. Hansmann, The Role of Non Profit Enterprise', Yale Law Journal 89, 1979-80, p. 835.
82. H. Srikanth : State, NGO's and Urban Community Developmet, Social Action, A Quarterly Review of Social Trends, A Social Action Trust Publication, Jan-March 1996, Vol 46, No. 1, pp. 42-45.
83. Pathak H., Citizen and Public Administration, in The Indian Journal of Public Administration, vol. XXI, No.3 II, Special Number Citizen and Administration II, New Delhi, 11PA, July - September, 1975, P.575.
84. Hulme, David and Paul Mosely, 1996, Finance Against Poverty, London, Routledge.
85. Hale S.M., 1983, "Decision process in Rural Development in India." Lucknow GFCS.

86. Harsh Sethi, Groups in a New Politics of Transformation', No. 7.
87. H. J. Eysenck, et. al., Encyclopaedia of Psychology, vol.2, London, Search Press, 1972.
88. Hoshiar Singh - Administration of Rural Development in India: Sterling Publishers Pvt. Ltd. New Delhi 1996.
89. Hoshiar Singh : Rural Development, Concept, Strategy and Approaches : Administration of Rural Development in India] (ed), Sterling Publishers, p. 3.
90. Hoshiar Singh : Rural Development Concept, Strategy and Approaches; Administration of Rural Development in India, Sterling Publishers, 1966. p.10.
91. H.J. Eysenck, et al., Encyclopaedia of Psychology, Vol 2, London, Search Press, 1972.
92. Hinchaliffe et al (Eds.)] nd : Fertile Ground: The Impacts of Participatory Watershed Management, Intermediate Technology Publications, London.
93. I.S. Hooda : Rural Development' Some Conceptual Issues; Journal of Rural Reconstruction, 29 (1) 1996, p. 1.
94. 27. Inayatullah, Approaches to Rural Development - Some Asian Experiences, Kuala Lumpur, Asian Center of Development Administration 1979.
95. Inayatullah, Approaches to Rural Development - Some Asian Experiences; Kuala Lumpur, Asian Centre of Development Administration, 1979.
96. J.M. Cohen and N.T. Uphoff, Participation's Place in Rural Development : Seeking Clarity Through Specificity, World Development, Vol 8, No. 30, 1980, pp. 213-216.
97. Juluis K. Nyerere, on Rural Development Impact, Vol 15, No. 3, March, 1980. p. 88.
98. J. M. Cohen and N. T. Uphoff, "Participation's Place in Rural Development: seeking Clarity Through Specificity, World Development, col.8, No.30, 1980, pp. 213 - 36.
99. J. M. Gandhi and Yugandhar Manadvkar : Involving Communism in Watershed Management Challenges Approaches : Paper Submitted to the International Seminar on "An Integrated Approach for Strengthening and Protecting Water Sources" at Pune on Sept. 27-28, 2001.
100. J. P. S. Ahuja & M. R. Rawtani - web-based information for Agriculture and Rural Development, University News, A weekly journal of higher education: Vol.21, May 27- June 2, 2002.
101. J.B. Kriplani, Gandhi - His Life and Thought, New Delhi, Publications Divisions, Ministry of Information and Broadcasting, Government of India] second reprint 1975 pp. 377-378.
102. Jathar and Beri, Indian Economics, Vols 1 and 2] Oxford University Press D.R. Gadgil Industrial Evolution in India] O.U.P. N.R. Inamdar] (ed) Political thought and Leadership of Lokmanya Tilak, New Delhi, Concept 1985; Vide Inamdar's Introduction and Bureaucracy and Political Ideology of Lokmanya Tilak'.

103. James H. Crops, Rural Sociology and Rural Development', Rural Sociology, Vol 37, No. 4, Dec. 1972, p. 515.

104. Joshi S. et al] (ed) 1997, Experiences of Advocacy in Environment and Development, Banglore, DSI and NOVIB.

105. John M. Cohen and Norman T. Uphoff : Rural Development Participation Concepts and Measures for Project Design, Implementation and Evaluation, Ithica] Cornell University, 1977 p. 7.

106. Kanchan Chopra, Gopal K. Kadekodi, M.N. Murthy : Participatory Development, Emergence of Participatory Institutions : Conclusions and Policy, Chap. 7, p. 139.

107. K.R. Sastri, Factor Affecting Rural Development Participation with Special Reference to IRDP and NREP, Journal of Rural Development, Vol 9, No. 6, 1990. p. 1078.

108. Kolavalli, Shashi and Jeff Brewer] 1999 : Facilitating user participation in Irrigation Management Irrigation and Drainage Systems Volume 13, pp 249 -273.

109. Kanchan Chopra, Gopal K. Kadekodi, M.N. Murthy : A Fresh Look at Rural Development and Strategies, Participatory Development : An Inroduction : Sage Publications - New Delhi. pp. 139-141.

110. Kothai, Krishna, 1994, Economic Impact of Rural Development Projects on Small Farmers, Journal of Rural Development, 13 (2).

111. Kumar R. Gopa, 1999, Foreign Contributions to Voluntary Organisations in India: A Geographical Analysis, JNU, New Delhi,] mimco .

112. Kanchan Chopra : Gopal K. Kadekodi, M.N. Murthy, Participatory Development : An Inroduction-A Fresh Look at Rural Development Strategies. pp. 17-18.

113. K. Finsterbusch and W. A. Van Wicklin III, " The Contribution of Beneficiary Participation to Development Project Effectiveness", in Public Administration and Development, New York, John Wiley, 1987.

114. Ker, John, N.K. Sanghi, and G. Sriramappa] 1999 : "Subsidies in Watershed Development Projects in India: Distortions and Opportunities, in Fiona Hincheliffe, John Thompson, Jules Pretty, Irene Guijt and Paramesh Shah,] eds Fertile ground: The Impacts of Participatory Watershed Management, IT Publications, London.

115. Kishore Chandra Padhy, Rural Development in Modern India] New Delhi, B.R. Publishing Corporation, 1986 P. 89-96.

116. Kishore Chandra Padhy, Rural Development in Modern India, p. 124.

117. K.M. Sen, Hinduism'] Baltimore : Penguin Books (1961) P. 23-24.

118. Khanna B.S. - Rural Development in South Asia - India, Deep & Deep Publications, New Delhi. 1991.

119. Lanita Charles : Deliberations on Publications of Development, ISPCK, Delhi,1986

120. Longwe S. 1990 from Welfare to Empowerment, a Post UN Women's Decade Update and Future, Directions, Working Paper 204, Michigan State University.

121. Leach Melissa, Robin Mearns and Jan Scoanes] 1999 : Environmental Entitlements: Dynamics and Institutions in Community Based Natural Resources Management', World Development vol.27, No.2, PP.225 - 247.

122. Leaf, M. J.] 1988 : Measures to Activate Farmers' Organisations, Louis Berger International, Inc and Water and Power Consultancy Services] India Ltd.] draft , New Delhi.

123. Long Norman, 1977 : An Introduction to the Sociology of Rural Development, London, Tavistock.

124. Leslie J. Calman, Toward Empowerment : Women and Movement Politics in India] Boulder, Co : West View, 1992 P. 21.

125. M. L. Santharam, et. at., "Herman and Social Factors in People's Participation", Journal of Rural Development, vol.1, No.5, 1982.

126. Madhu Mishra : Rural Development-Action and Management] (ed) P.61-65.

127. M.K. Gandhi, Hind Swaraj or Indian Home Rule, Ahmedabad, Navjivan Publishing House,] revised edn (1939). Reprint 1982] first edn. 1938 P. 104.

128. L. Mishra, 1994. The Anguish of the Deprived, Delhi. Har-Anand.

129. M.K. Dubey - Rural and Urban Development in India. Common Wealth Publishers, Delhi 2000.

130. Maheshwari S.R. 1985, Rural Development in India, Delhi, Sage.

131. M.K. Dubey, Rural and Urban Development in India : Centrally Sponsored Schemes] (ed) Common Wealth 2000, P. 267-268.

132. Murthy R.K. and Nitya Rao, 1997, op.cit, P. 57.

133. M.L. Santhanam : Community Participation for Sustainable Development, The Indian General of Public Administration, Special Issue on; Towards Sustainable Developmet of Society - Imperative and perspectives, July-Sept. 1993 Vol XXXIX, No : 3] (ed) P. 413.

134. Marilyn Gittel, Limits to Citizen Participation', Sage Library of Social Sciences, 109, P. 22-23.

135. Mishra B.] 2996 : A successful case of participatory watershed management at Ralegaon Siddhi Village in District Ahmednagar, Maharashtra.

136. Michael Banton, "Anthropolitical Aspects, Voluntary Associations" in David Sills] (ed) International Encyclopaedia of Social Science, Vol 16, New York. The Macmillan Company and the free press, London, Collier Macmillan 1968. P. 357.

137. M. L. SANTHANAM: Community participation for sustainable development, P.413, The Indian Journal of Public Administration, Special issue on; Towards sustainable development of society - imperative and perspectives, July - September 1993, vol. XXXIX, Number 3] ed. T. N. Chaturvedi.

138. Milton J. Esman, "The Politics of Development" in John D. Montgomery and William J Siffin] eds Approaches to Development, New York. McGraw Hill. 1966 PP.59 - 112.

139. M.L. Dantwal - Search for Employment oriented growth Strategy. EPW. May 26th, 1990.

140. M. Laxmi Narasaiah and G. Jaya Raju : Rural Development and Anti-poverty programme. Discovery Publishing House, New Delhi 1999.

141. M.A. Muttalib, Voluntarism and Development - Theoretical Perspectives : The Indian Journal of Public Administration Quarterly Journal : Special Number on Voluntary Organisations and Development, Their Role and Functions, July-September 1987, Vol XXXIII, No. 3.

142. Murthy R.K. and Nitya Rao, 1997 : Indian NGO's, Poverty Alleviation and their Capacity Enhancement in the 1990's : An Institutional and Social Relations Perspective, New Delhi, FES.

143. Mohit Bhattacharya, Bureaucracy and Development Administration, New Delhi, Uppal Publishing House, 1991. P. 88.

144. Moore M., 1993, Good Government? Introduction IDS Bulletin, 24] 1 .

145. Murthy R. K., and Nitya Rao, 1997, Indian NGOs, Poverty Alleviation and their Capacity Enhancement in the 1990s : An Institutional and Social Relations Perspective, New Delhi, FES.

146. Namerta, 1999, Income and Employment Generation Programmes of NGOs in India, Bangalore: Development Support Initiative, Forthcoming.

147. Namrata, 1999 : Income and Employment Generation Programmes of NGO's in India, Banglore : Development Support Initiatives, Forth Coming.

148. N.C. Saxena, Instances of some Anti-poor government Policies in India.' Paper presented at the UNDP Seminar on " Poverty Eradication - the potential of Community Empowerment! Delhi 17 - 18 October 1996.

149. Norman Johnson, Voluntary Social Services, Oxford, Basil Blackwell and Martin, Robertson, 1981, P. 14.

150. Norman Uphoff, Fitting Projects to people' in Cernea] (ed) Putting People First : Sociological Variables in Rural Development, New York : Oxford University Presss, 1985. P. 381.

151. N.R. Inamdar : Role of Voluntarism in Development, The Indian Journal of Public Administration Quarterly Journal 6. (1961). P. 129.

152. Nadeem Mohsin, Rural Development Through Government Programme,] Delhi, Mittal Publications, 1985 , P. 15-16.

153. Oley S.K.] 1985 Voluntary Action or Collusion ? Kurukshetra Vol 34, No. 1, Oct. 1985, P. 5.

154. Pathak H., Citizen and Public Administration in the Indian Journal of Public Administration, Vol XXI, No : 3 II, Special Number - Citizen and Administration II, New Delhi IIPA, July-Sept. 1975. P. 575.

155. Pillai J.K, 1995, Women and Empowerment, Gyan Publishing House, New Delhi.

156. Pillai J.K, 1999, Empowering Women in India: New roles for Education. Education and Women Empowerment, University news, Association of India University, New Delhi.

157. Pillai J.K. : Women and Empowerment] (ed) 1995, Gyan Publishing House, New Delhi, P. 38-40.

158. Prabhakar Singh, Community Development Programmes in India, New Delhi, Deep and Deep Publications. 1982. P.34

159. Prabhakar Singh, Community Development Programme in India, New Delhi, Deep and Deep Publication, 1982, P. 34.

160. Piciotto, Robert] 1992 : Participatory Development - Myths and Dilemmas, WPS 930, The World Bank, Washington.

161. Pretty Jules, N] 1995 : Participatory Learning for Sustainable Agriculture, World Development, vol. 28, No.8, PP. 1247 - 1263.

162. P. Oakely et. al., Projects with people: The practice of participation in Rural Development, Geneva , ILO, 1991.

163. Piciotto, Robert] 1992 : Participatory Development: Myths and Dilemmas, WPS 930, The World Bank, Washington.

164. Peter B. Clark and James Q. Wilson, Incentive Systems : A Theory of Organisations.' Administrative Science Quarterly 6, (1961), P. 129-66.

165. Ramesh Vaswani., Micro-Watershed Development., Three success Stories from Maharashtra., Published by Yeshwantrao Chavan Academy of Development Administration 1995, P. 50.

166. R.T. Tiwari and R.C. Sinha, Rural Development in India, Ashish Publishing House, New Delhi. 1998. P.1

167. R.P. Mishra and K.V. Sunderam, Rural Development perspectives and Approaches. New Delhi., Sterling Publishers, 1979, P.4

168. Rao K.B. 1999. Safe Motherhood. In S.S. Ratnam] Eds Obstetrics and Gynacology of Post Graduates vol.1, 2nd Edn. Chennai - Orient Longman P. 1-7.

169. Richard Taylor, The Encyclopaedia of Philosophy,] Paul Edwards, Editor-in-Chief Vol. 8 New York, The Macmillan Company and the Free Press London, Collier Macmillan 1967 P. 270.

170. Robert L. Hardgrave, Jr. and Stanley. A. Kochank, India : Government Politics in a developing Nation] San Diego : Har Court Brace, Jovanovich Publishers, 1986 P. 52.

171. Robert Chambers, Rural Development : Putting the Last First, London Longman, 1983, P. 147.

172. Robert Chambers, Rural Development : Putting the Last First] London : Longman, 1983 .

173. Robin William, A. The Civil Service in Britain and France, London, The Hogarth Press, 1956.

174. Riddel R. and Mark Robinson, 1992 : The Impact of NGO Poverty Alleviation Projects : Results of Case Study Evaluations, Working Paper No. 68, London, Overseas Development Institute.

175. Ramesh Arora : Summing Up' in Rakesh K. Arora] (ed) Politics and Administration in Changing Societies : Essays in Honour of Professor Fred W. Riggs, Op at, P. 350.

176. Rohini Patel, op.cit, P. 50-51.

177. Robinson M. 1993, Governance, Democracy and Conditionality : NGO's and the New Policy Agenda in A. Clayton] (ed), Governance, Democracy and Conditionality : What role for NGO's ? Oxford, INTRAC.

178. Ramola Baxamusa and Hema Subramaniam, 1992 : Assistance for Women's Development from National Agencies : Employment Programmes, Mumbai, Popular Prakashan, P. 70.

179. Robinson William A., The civil Service in Britain and France, London, The Hogarth Press, 1956.

180. Rohini Patel, Voluntary Organisations in India : Motivations and Roles', in Social Change Through Voluntary Action] (ed) Chap. 3. P. 41.

181. Rajasekhar D., 1996, Problems and Prospects of Group Lending in NGO Credit Programmes in India, Saving and Development, 20] 1 .

182. Reddy N.L., Narsinha and D. Rajasekhar, 1996, Development Programmes and NGOs : A guide on Central Government Programmes for NGOs in Indial, Bangalore, BCO and NOVIB.

183. Riddel R., and Mark Robinson, 1992, The Impact of NGO Poverty Alleviation Projects: Results of the Case Study Evaluations, Working Paper No.68, London, Overseas Development Institute.

184. Robinson M., 1993, Governance, Democracy and Conditionality: NGOs and the New Policy Agenda, in A. Clayton] ed.,, Governance, Democracy and Conditionality: What Role for NGOs? Oxford, INTRAC.

185. Saangitha S. N., 1990, Self-Employment Programme for Rural Youth: The Role of Non-Governmental Organisations] NGOs , IIMB Management Review,5] 2 .

186. Stephen F., 1990, NGOs Hope of the Last Decade of this Century, Bangalore, SEARCH.

187. Sundaram I.S., 1986, Voluntary Agencies and Rural Development, New Delhi, B.R. Publishing Corporation.

188. Subrata Kumar Mitra, Crisis and Resilience in Indian Democracy.' International Social Sciences Journal 43, 1991. P. 565.

189. S.K. Chatterjee : Concepts of Development and Modernisation : Development Administration, Chap. 4, P. 34-35.

190. See R. Hooja, The District as a Planning Unit, Style and Locus' and N.R. Inamdar, District Planning in Maharashtra' both in Indian Journal of Public Administration, XIX] July-Sept. 1974 , P. 393-406, 320-27, Sudipo Mundle, District Planning in India, Delhi, Chapter II.

191. S. Nagendra Ambedkar : People's Participation in Rural Development. P. 98-99.

192. S.N. Ambedkar, In Rural Development Programme : Implementation Process] Jaipur, Rawat, 1994

193. Swapan Garain : Government-NGO Interface in India - An overview, Indian Journal of Social Work, Vol No. 3,] July 1994 .

194. Sanjit] Bunker Roy : Voluntary Agencies in Development : The Indian Journal of Public Administration, Quarterly Journal of the Indian Institute of Public Administration, Special Number on Voluntary Organisation and Development, their Role and Function. P. 461.

195. S. Paul, "Community Participation in Development Projects: The World Bank Experience", in Readings in Community Participation, Washington, D. C. The World Bank, 1987.

196. Sachchidananda, 1998. Social Change in Village India. Delhi, Concept Publishing House.

197. Sachchidananda, 1998. Social Change in Village India. Delhi, Concept Publishing House.

198. Sangitha S.N. 1990 : Self Employment Programme for Rural Youth : The role of Non-Governmental Organisations] NGO's IIMB Management Review, 5] 2 .

199. S.C. Jain, S.K. Verma, Manisha Khale, Nitin Inamdar : Final Evaluation Report, Participatory Watershed Development Programme] 1996-2002 .

200. S.C. Jain, Community Development and Panchayati Raj in India, New Delhi, Allied Publishers, 1967, P. 56.

201. Sinha M.S.] 1995 : Review and Status of Rainfed Farming and Watershed Management in India in P.N. Sharma and M.P. Wagley] eds . The Status of Watershed Management in Asia, UNDP/FAO Katmandu, Nepal.

202. Seth, S.L.] 1996 : The National Watershed Development Programme For Rainfed Areas] NWDPRA : Restrospect and prospects' in Jensen, TR, et al] eds Watershed Development - Emerging Issues and Framework for Action Plan for strengthening a learning process at all leve, proceedings of DANIDA'S International Workshop on Watershed Development, Hubli and Bangalore, India, December 1995, WD CU, New Delhi

203. Sitakanta Sethi, Participatory Planning for Rural Development, Ashwattha-Quarterly Journal of YASHADA, Vol 3, No. 1-3, Jan.-Sept. 2000. P. 16-18.

204. Salahuddin Md. Aminuzzaman, " Community Development to Integrated Rural Development : The Context of Asia Pacific Region." South Asian Studies vol. 19, No. 1] Jan to June 1984

205. S.N. Mishra, New Horizons in Rural Development Administration] Delhi - Mittal Publications, 1989 P. 21

206. Seshadri, Procedures of Planning in Developing Countries. 1970 . P.3

207. S.S. Gadkari, A.S. Kharkar, V.V. Velankar : Organisation of the State Government in Maharashtra, Himalaya Publishing House. Bombay.

208. Shriram Maheshwari, Rural Development in India : A Public Policy Approach] New Delhi : Sage Publications, 1985 , P. 13.

209. Shriram Maheshwari : Voluntary Action in Rural Development in India, Quarterly Journal of the Indian Institute of Public Administration, July-Sept 1987, No. 3, Vol XXXIII. P. 560.

210. Sinha M. S.] 1995 : Review and Status of Rainfed Farming and Watershed Management in India' in P. N. Sharma and M. P. Wagley] (Eds.). The status of Watershed Management in Asia, UNDP/FAO Katmandu, Nepal.

211. Seth, S. L.] 1996 : The National Watershed Development Programme For Rainfed Areas] NWDPRA : Retrospect and prospects' in Jensen, TR, et al] (Eds.) Watershed Development - Emerging Issues and Framework for Action Plan for strengthening a learning process at all level, proceedings of DANIDA'S International Workshop on Watershed Development, Hubli and Bangalore, India, December 1995, WDCU, New Delhi.

212. S.K. Sharma and S.L Malhotra, Integrated Rural Development. Approaches, Strategy and perspectives, Abhinav Publications, New Delhi, 1977, P.16-17

213. Sundaram I.S - Anti-Poverty Rural Development in India : P.K. Publications, New Delhi. 1984. P. 181 - 182

214. S.K. Singh : Strategies for Rural Development : An Overview, P. 72-73.

215. S.N. Mishra, New Horizons in Rural Development Administration] Delhi, Mittal Publications, 1989 , P. 2.

216. S.T. Kachwe, G.K. Sangle and V.K. Patil; Opportunities for Marginal and Landless Labourers in Drought Area of Marathwada; Rural India, January 1981.

217. S. Paul, Community Participation in Development Projects, The World Bank Experience' in Readings in Community Participation Washington D.C. The World Bank, 1987.

218. Sakuntala Narasimhan : An Overview of Past State Initiatives, Empowering Women] (ed), Sage Publications, New Delhi, P. 36-37.

219. Turton, Cathryn, John Farrington, and John Coultier] 1998 : A review of the Implantation of the 1994 Watershed Management Guidelines] Draft , Overseas Development Institute, and London.

220. T.N. Dhar, NGO's : Their Roles, Responsibilities and Problems; Quarterly Newsletter, Dynamic Administration] (ed) S.P. Gupta, U.P. Regional Branch : Indian Institute of Public Administration] Jan-March 2000 , 49th Issue.

221. Uma Lata, She Design of Rural Development - Lesson from America, Baltimore : John Hopkins University Press, 1975, P. 20.

222. Uma Lele, The Design for Rural Development : Lessons for Africa, Baltimore, John Hepkins, 1975.

223. Uma Lele, The Design for Rural Development: Lessons for Africa, Baltimore, John Hepkins, 1975.

224. Vasant Desai, Rural Development, vol.1 Issues and Problems] Bombay : Himalaya Publishing House 1988 P. 380

225. V.M. Dandekar and N Rath, Poverty in India. Allied Publishers, New Delhi. 1970. P.12

226. V.H. Desai, The Visionary, Important Speeches and Writings of Swami Ramanand Teerth, Published : Swami Ramanand Teerth', Marathwada University, Nanded, 1996. P. XX.

227. V. Sivalinga Prasad, K. Murali Manohar : Fred W. Riggs : Administrative Thinkers

228. Vasant Desai, Rural Development; Issues and Problems, Vol I, P. 381.

229. Vasant Desai, Rural Development; Programmes and Strategies, Vol II, P. 88-89.

230. Vasant Desai, Rural Development, Rural Development Through the Plans, Vol V, P. 47.

231. V. Mohini Giri : Forward - Empowering Women; An Alternative Strategy from Rural India in Sakuntla Narasimhan, Sage Publications, New Delhi, 1998, P. 10.

232. Vaidyanathan] 1991 : Integrated Watershed Development: Some Major Issues', Foundation Day Lecture, Society For Promotion of Wasteland Development, New Delhi.

233. Verma H.S. 1975, " A Critical Appraisal of Community Development programme in India." IIM, Ahmedabad.

234. Vinoba Bhave, Introduction in K.G. Mashurwala, Gandhi and Marx', Ahmedabad Navjivan Publishing House, 1951. P. 24-26.

235. V.T. Krishnamachari, Community Development in India', New Delhi, Government of India, Publication Division, 1958, P. 12.

236. Webster, Neil, 1995 : The role of NGO's in Indian Rural Development : Some lessons from West Bengal and Karnataka, The European Journal of Development Research 7] 2 .

237. Webster, Neil, 1995, The Role of NGOs in Indian Rural Development: Some Lessons from West Bengal and Karnataka, The European Journal of Development Research, 7] 2 .

238. William Beveridge, Voluntary Action in a changing world, London, Bedford Square Press, National Council of Social Services, 1979. P. 1-5.

239. William Beveridge, op.cit, P. 100.

240. Yugandhar Mandavkar, Equity in Watershed Development, Paper Presented at The Workshop on Participatory Watershed Development Organised by Swiss Agency per Development and Co-operations] SDC Project.

241. Yugandhar Mandavkar - Information on Marathwada Sheti Sahayya Mandal, December 1997.

242. Yugandhar Mandavkar, Case Study on Gender and Watershed Management' - Synopsis, Paper Presented for Gender and Technology Workshop organised by Anthra at Hyderabad on July 27-30, 2001.

243. Yugandhar Mandavkar., Equity in Watershed Development., A Case Study on Adgaon Project., Presented at the Workshop on Participatory Watershed Development organised by Swees Agency for Development & Co-operation's] SDC Project Support Centre for the Partners of Indo-Swees Participatory Watershed Development Project - Karnataka] ISPWD-K at National Institute of Agricultural Extension Management] MANAGE ., Hyderabad on September 2-4, 1998.

244. Yugandhar Mandavkar., Watershed Development : Shifts in Involving Women., Paper presented at the Workshop on "Women's Involvement in Watershed Development" organised by People's Action for Watershed Development Initiatives] PAWDI project at Jaipur on July 29 - August 1, 1996.

245. Y.N. Rao, Administration of Rural Development Programme' : A Study in Ranga Reddy District of Andhra Pradesh, Journal of Rural Development, Vol 9, No. 6, P. 990.

Articles on Voluntary Agencies and Rural Development

1. " Promoting rural development through voluntary action." : Interview with Shri Rangan Dutta, DG, CAPART Kurukshetra, 49] 8 , 2001] May : 2-11.
2. Barik B.C. : Voluntary Organization, lift irrigation and rural development. Indian Economic Panorama, 4] 1 , 1994] April : 35-43.
3. Bebbington, A. : New States, new NGOs ? Crisis and transitions among rural development NGOs in the Andean Region. World Development, 25] 11 , 1997] November : 1755-1767.
4. Bhaumik , Alok K. : People-Centred NGO Initiative in Sustainable Rural Development: Case of Rangabelia Project in West Bengal] India . Indian Journal of Regional Science, 28] 2 , 1996 : 111-118.
5. Egger, Philippe : Rural Organisations and infrastructure projects: Social investment comes before material investments. International Labour Review, 1] 131 , 1992 : 45-62.
6. Francis, C. : Rural Development, people's participation and the role of NGOs. Journal of Rural Development, 12] 2 , 1993] March : 205-210.
7. Ghosh D.K. : Voluntary Agencies and rural uplift, Yojna, 34] 18 , 1990] October 1-15 : 13-14 and 26.
8. Gopalakrishna K. : Alleviating rural poverty through voluntary efforts : A Study, Kurukshetra, 39] 5 , 1991] February : 26-28.
9. Hooja B. : Towards Sadabahar Vikas' Sustainable development, Prasar, 20] 1-2 , 1993] April-June and July-Sept. : 3-11.
10. Kapoor R.N. : Voluntary Organization in promoting science and technology in rural development, Manthan, 12] 1 , 1991] January : 9-14.
11. Kothai, Krishna : Voluntary action and rural development. Indian Economic Panorama, 5] 4 , 1996] Jan-March : 22-25.

12. Mahipal : Role of voluntary organizations in rural development. Social Changes, 21] 3 , 1991] September : 24-30.
13. Mahipal : Role of NGOs/Voluntary organizations in rural development. Social Changes, 21] 3 , 1991] September : 24-30.
14. Mehrotra S.C. : Role of NGOs/Voluntary organisations in rural development. National Bank News Review, 6] 8 , 1990] October : 32-34.
15. Nanda, Nitya : Emeging role of NGOs in rural development of India. Social Changes, 30] 3/4 , 2000] Sept-Dec : 34-41.
16. National approach to social development, Yojna, 39] 6 , 1995] April : 6-9.
17. Rajasekhar D. : Rural development strategies NGOs, Journal of Social and Economic Development, 1] 2 , 1998] Jul- Dec : 306-327.
18. Sachidananda : Role of Voluntary agencies in tribal areas, Lokayan Bulletin, 10] 5-6 , 1994] March-June : 91-98.
19. Samhungu, Farai and Masendiki, Absolom Scaling up : growing up with development, Appropriate Technology, 24] 3 , 1997] December : 1-3.
20. Scoones, Ian : Replicating islands of success, Appropriate Technology, 24] 3 , 1997] December .
21. Sesay, Sullay B. : Non-governmental organizations] NGOs and rural development in Sierra Leone, Adult Education and Development, 48, 1997: 245-254.
22. Shanta Kohli, Chandra : NGOs and women's development in rural South India: A comparative analysis] Vanita Viswanath , Indian Journal of Public Administration, 39] 4 , 1993] Oct-Dec : 714-715.
23. Shanthi K. : Role of voluntary organizations in rural development in the 21st century: Issues and challenges, Political Economy Journal of India, 6] 1 and 2 , 1998] Jan-June : 75-85.
24. Shripathi K.P. : Performance of Rural Development Projects by Voluntary Agencies, Journal of Rural Development, 14] 4 , 1995] Oct-Dec : 431-442.
25. Singaiah, G. : Managing rural development - NGOs strategies, Indian Economic Panorama, 7] 1 , 1997] April-June : 47-48.
26. Srivastava Alka : Non-government organizations and rural development, Social Action, 49] 1 , 1999] Jan-Mar : 27-44.
27. Shrivastava I. C. : NGOs Role : Micro-level perception, National Bank News Review, 11] 4 , 1995] Oct-Dec : 10-13.
28. Suresh K.A and Joseph, Molly : Peoples participation in rural development a comparative study of co-operatives and voluntary organisations, Co-operative Perspective, 25] 1 , 1990] April-June : 46-50.
29. Uphoff, Nerard : Grassroots organization and NGOs in rural development : Opportunities with diminishing states and expanding markets, World Development, 21] 4 , 1993] April : 607-622.

APPENDIX - I

List of Abbreviations and Acronyms

- ACO - Area Community Organiser
- AFARM Action for Agricultural Renewal in Maharashtra
- AFPRO- Action for Food Production
- AIDS - Acquired Immune Deficiency syndrome
- AIWC - All India Women Conference
- AKF - Aga Khan Foundation
- ARWSP Accelerated Rural Water Supply Programme
- ASCS - Agricultural Service Centers
- AVARD Association of Voluntary Agencies for Rural
- BNGOs Business Sponsored NGOs
- CAP - Community Action Programme
- CAPART Council for Advancements of peoples Action and
- CAT - Centre for Appropriate Technology
- CATL - Co-operative f+or American Relief Everywhere
- CBOs - Community Based Organisations
- CIBA - Canadian International Development Agency
- CO - Community Organiser
- CRM - Vommunity Resource Management
- CRSP - Central Rural Sanitation Programme
- CRY - Child Relief and You
- DA - Developmental Alternatives
- DANIDA Danish International Development Agency

- DDA - Danish Development Agency
- DDP - Desert Development Programme
- DISP - Development Initiative Support Programme
- DOs - Development Organisations
- DPAP - Drought Prone Area Programme
- DWCRA Development of Women and Children in Rural Area
- E D I - Enterprise Development Initiative
- EAS - Employment Assurance Scheme
- EFCD - Employment For Community Development
- ERM - Environment resource Management
- EZE - Evangelische Zentralastelle Fur Entwick Lungshelfe V
- FEVORD Federation of Voluntary Organisations in Karnataka
- FPCs - Forest Protection Committees
- FUNGOs Donors-sponsored organisations
- GIT - Geohydrological Investigation Team
- GNP - Gross National Product
- GOI - Government of India
- Gongos Government sponsored non-governmental organisations
- GOs - Government Organisations
- GRACE Guild of Regional Associates for Community Enpowerment
- GRASP- Grass-roots Action for Social Participation
- GROs - Grass-roots Organisations
- GTA - German Agency for Technical Assisting
- GTZ - German Agency for Technical Assistance
- HIV - Human Immuno-deficiency Virus
- I L O - International Labour Organisation
- IAY - Indira Awaas Yojana
- IDE - International Development Enterprises
- IGWDP Indo-German Watershed Development Program
- IRDP - Integrated Rural Development Programme
- ISB - Industry Service Business
- IT - Intermediate Technology
- IWDP - Integrated Watersheds Development Project Scheme
- IYN - International Year of Volunteers

- JFM - Joint Forest Management
- JRY - Jawahar Rozgar Yojana
- KP - Krushak Panchayat
- KPL - Karnataka Pulpwood Limited
- LIST - Local Initiative Support Team
- Manavlok Marathwada Navnirman Lokayat
- MIR - Mid-term Review
- MMs - Mahila Mandals
- MOA - Ministry of Agriculture
- MORD - Ministry of Rural Development
- MSSM - Marathwada Sheti Sahayya Mandal
- MSW - Million wells schsemes
- MYRADA Mysore Resettlement Agricultural Development Agency
- NABARD National Bank for Rural Development
- NADWM National Drinking Water Mission
- NCF - National Children's Fund
- NGDOs Non-Governmental Development Organisations
- NGOs - Non-Governmental Organisations
- NIPCCD National Institution of Public cooperation and child
- NIRD - National Institute of Rural Development
- NOVIB - Netherlands International Development Cooperation
- NREP - National Rural Employment Program
- NWDB - The National Watershed Development Board
- NWDPRA National Watershed Development for Rainfed Agriculture
- OBAP- Organization Of Beneficiaries Of Anti-Poverty Programme
- OXFAM International Development Aid Agency
- PADI - Programmes by Action In Development India
- PCO - Programme for Community Organisation
- PNGOs Political party sponsored organisations
- PLA - Participatory Learning and Action
- PRA - Participatory Rapid Appraisal
- PRA - Participatory Rural Appraisal
- PRIA - Society for Participatory Research in Asia
- PRIs - Panchayat Raj Institutions

- PVARD Promotion For Voluntary Action In Rural Development
- Quangos Quasi autonomous NGOs
- RGNDWM Rajiv Gandhi National Drinking Water Mission
- RLEGP - Rural Landless Employment Guarantee Programme
- RWSP - Rural Water Supply Programme
- SDC - Swedish Development Co-operation
- SEWA - Self-employed Women's Associations
- SHG - Self - Help Group
- SI - Sector In-Charge
- SIDA - Swedish International Development Agency
- SMI - Safe Motherhood Initiative
- SMOs - Social Movement Organisations
- TREAD Team for Research, Evaluation and Development
- TRYSEM Training of Rural Youth for Self-employment
- UNESCO United Nations Educational, Social and Cultural
- VA - Voluntary Agency
- VANI - Voluntary Action Network India
- VDO - Voluntary Development Agencies
- VHAI - Voluntary Health Association of India
- VITA - Volunteers In Technical Assistance
- VLW - Village Level Workers
- VOLAGs Voluntary Agencies
- VOLOGs Voluntary Organisations
- VOs - Voluntary Organisations
- VWCs - Village Watershed Committees
- WDP - Watershed Development Program
- WOTR - Watershed Organisation Trust
- YIP - Young India Project
- YMCA - Young Men Christian Association

APPENDIX - II

Interviews of the Experts working in the field of

Voluntary Organizations and Government

Dr. Ghare

The secretary, AFARM, Action for Agricultural Renewal in Maharashtra. He stated that the NGOs are doing their best in all the sectors and especially in the watershed project. He said that watershed program is a new participation concept, where NGOs have sought the involvement of the people in planning process and the people have shown full commitment for the first time to upgrade the human resources. Dr. Ghare said that this community participation is a value addition for the life of people. Agriculture being the main economy source, people can realize the mutual benefit of watershed program, by involving themselves in the planning as well as implementation process. He said these watershed project definitely relates to generation of wealth thereby increasing the living standard of the rural people.

Shri Ramesh Vasvani

He works for Yeshwantrao Chavan Development Academy (YASHADA). Shri Vasvani has a vast experience as researcher and teacher with different semi-government and non-government organizations on various tribal, rural and urban development projects sponsored by government of Maharashtra, India and international government and NGOs. Shri Vasvani already has a couple of books and few short articles to his credit. In one of his books; Micro-Watershed Development' has appreciated the work of Naigaon, Adgaon and Ralegan Siddhi.

Discussing the importance of voluntary agency in rural development Shri Vasvani stated that government organizations are over-burdened with its routine administration and cannot concentrate on developmental activities. The projects organized by government lack flexibility. He added that the administrative machinery at village level is less educated as well as less in number to carry out the programs at local level so are inefficient to co-ordinate people along with minimum funds at their disposal. Shri Vasvani commented that voluntary agencies have flexibility of approach, which is the only method to bring about integrated development. He firmly stated that NGOs have dynamic leadership together with supplementary funds.

Shri Vasvani positively said that the NGOs could act as nodal agency to implement projects and bring about community participation for integrated development. He added that the government should promote such NGOs by instituting awards. The government can also grant recognition to such NGOs, along with organization of technical and administrative training programs for their employees and leaders. Shri Vasvani firmly stated that the government should either take help of the NGOs in the process of rural development or help the NGOs in planning, implementing and supplementary funding. With reference to the projects of Naigaon, Adgaon and Ralegan Siddhi, Shri Vasvani commented that the government agencies extended full support, which resulted in successful implementation of the work.

Turning towards the Adgaon project, Shri Vasvani asserted that, YASHADA had learnt the success of watershed project in Adgaon, because Adgaon is primarily situated in the drought-prone region, where irrigation is very difficult. He added that Adgaon being the last project among the three it had the advantages to borrow and errors to be over-looked. He said MSSM had well-planned project proposal together with comprehensive surveys and systematic involvement of the people. He again added that, MSSM involved the government agencies efficiently in the project viz; Minor irrigation Department, Soil Conservation And Agriculture Department, Forest department etc. Shri Vasvani appreciated the agencies project implementation and administration. He mentioned that the credit of the fieldwork goes to Shri Vijayanna Borade and the planning applause to hon. Shri. Gandhi. He remarked that the agency has motivated leaders, who also have exposure to urban areas and atmosphere, which made them apply the quality in their project. Shri Vasvani added that they carry vast experience of higher level technology, which brought superiority in their work and commented that they have qualities of dedication and sacrifice.

Shri Vasvani also shared his views towards government agencies by saying that government has funds for such projects, and the projects like Adgaon devote full concentration on a particular village, co-ordinate and implement schemes.

Dr. Bhogle

The former secretary of Water and Land Management Institute [VALMI]. It was a useful discussion with him on NGOs in general. Regarding rural development, Dr. Bhogle explained that generally for India and in particular for Maharashtra the efforts of government as well as NGOs are necessary. He said they should cooperate with each other in the development projects and decide clearly on what they have to do. If the government is undertaking projects viz., health, water, education etc., NGOs should look after the maintenance. Dr. Bhogle mentioned that government agency is most of the time funding agency on the other hand NGOs are skill oriented organization with proper planning, participation, having knowledge of the problem etc. Adding to his talk Dr. Bhogle said that NGOs are non-profit making organizations and stick to the principles of work. While talking on the working method of NGOs Dr. Bhogle said they are flexible in their principles, there occurs no bossism in the employees, there's transparency in their work and very importantly their project cost is less.

Lastly Dr. Bhogle stated prominently that for the NGOs to remain working with same motto of selflessness it is necessary to reward them which can bring the sense of satisfaction, pride and appreciation.

Vilasrao Salunkhe

Shri Vilasrao began his discussion with the perspective of rural development. He said it is necessary to know the needs of the grass-root people and study their socio-economic standards. He very promptly stated that since independence government is unsuccessfully striving to implement the policies and programs for rural development. On the other hand large number of NGOs are well implementing and executing this development programs as the rural development is a community program.

Shri Vilasrao mentioned that Marathwada is basically rural area. He said though there is increase in population nationally, only 45% of the population resides in the urban areas increasing the population pressure in the rural area. He well said that the population in Adgaon itself doubled in last two years. All this enhances the administrative responsibilities of the government leaving very less of the time for development activities. Again the population expansion decreases the employment opportunities or the government is unable to create those. Along with it there is degradation of natural resources, water-level going down, cutting of forest, land degradation, due to all these reasons especially in rural areas it makes irrigation difficult, so no gainful employment for the people. Shri. Vilasrao stated that the outcome of this phenomenon is poverty, unemployment and degradation of natural resources so then there is the need of natural resource management.

Shri Vilasrao added that the government is bringing up number of development programs to handle the situation created above, but cannot make a sea change for the grass-root people to meet their needs. This failure of the government programs brought the NGOs to work in this direction viz., Harnessing natural resources, tree-plantation, watershed projects, dairy, soil conservation involvement of local people in executing the project work etc.

After sharing the views on NGOs Shri Vilasrao spoke on few problems regarding the State and NGO. He questioned how far the NGOs have been able to achieve integrated approach in total development. He argued that the NGOs have very well carried on the activities like economy, health, education, agriculture etc. but whether the projects are sustainable? have they really given development opportunity to everyone in the rural sector? are they able to therefore village structure?

After few of these queries Shri Vilasrao was positive in saying that the NGO is better than the State agency because it works with the concept of equity which should be adhered to achieve rural development for community as a whole and the State lacks in this principle.

Commenting on the future of India Shri Vilasrao stated that, apart the governments Family Planning Programs the population of the country would be increasing tremendously in next 20 years. This will heighten the gap between rural and urban development. To nurture this situation, rural development should not be down grading contrarily rural natural resources should be regenerated, the development programs must be sustainable and equity-based.

Turning towards development scenario after independence Shri Vilasrao said though land and water are social structures, the State has scarce water resources compared to the land resource, which should be developed. He remembered that the Agrarian Movement after the independence for the distribution of land was unsuccessful, and added that the distribution of land is impossible. Especially in Maharashtra the land is rain-fed so the land must be made productive. Shri Vilasrao trying to solve this problem further stated that water being a scarce resource should be accessed with equity by the State, people, area and the village. On concluding he said that, the rural people must be socially empowered to attain development.

Mr. Abraham

The Manager WOTR (Watershed Organisations Trust), shared valuable thoughts on the working of NGOs in rural development. Adding to the compliment of watershed perspective Mr. Abraham said that it can have excess of water and give good productivity in agricultural related activities. He said watershed can make great changes in rural sectors. He particularly mention that government should work through NGOs and NGOs should also try to work in collaboration with the government agencies. All the NGOs should have positive approach towards the government agency and he is happy that the government is recently lending their support to the NGOs.

Mr. Vidhate

The technical in-charge WOTR, also shared his experienced thoughts on the working of NGOs in rural development. He said the intervention of the NGOs in all the rural aspects, viz., health, education, watershed, awareness etc., is very necessary, as the government is not inclined to work at the grassroots. There is no transparency in the government working and so the local people never know what is the project about and how useful it would be for them. The GOs have no people's participation in planning and implementation of the projects and so the needs and aspirations remain neglected. He categorically mentioned that various programmes are implemented, but there is no follow-up as to evaluate the success or the failure of the programmes. He also said that, the government programmes are not target oriented.

Mr. Abraham and Mr. Vidhate adding to the overall perspective of Marathwada said that Marathwada has good soil and water level, but the politicisation of the working and implementing of projects in Marathwada than Western Maharashtra is hampering the development of rural sector. They also added that Western Maharashtra is better co-operative movement than Marathwada. In Marathwada there is no grassroot level leadership which lacks awareness regarding the government affairs, policies and lack of awareness for right to demand for their fundamental rights, help and needs. They said that Marathwada has a very low rate of literacy and no proper infra-structure. It has no dairy movement. There are very less number of credit co-operatives while the old ones are defunct. On asking about the working of MSSM in Marathwada, they said that, "MSSM is an ideal organisation and it influenced the government machinery with its project".

Both concluded by saying that Maharashtra had a good atmosphere of NGOs always, with traditional leadership of Phule, Shahu, Ambedkar and the State will also have a bright future with the NGOs.

Collector of Aurangabad, V. Radha

The Collector said that basis of rural development depends on the agricultural development and for the development of the rural sector the root cause has to be discovered. She feels that water' is the basic source for the overall development in the rural areas as people depend solely on agricultural farming and agricultural related business. So the focus of rural development basically should be water conservation and watershed. V. Radha said, a lot of good NGOs are doing tremendous work in rural development and watershed development. Their work is complimentary to the government organisations. She added that government has too many schemes for the rural poor which has been a mere failure, viz., well digging and construction, cow distribution, health programme etc. Instead the government should widen the agrarian base and concentrate more on infra-structure which will develop the area as whole and make the community economically stable. She said the present working of Aurangabad division of government organisation in rural sector is very good and government is doing a lot of work in the area of watershed project. She belives if the government officials / workers work hard and sincerely they can positively change the future of rural areas. She is also happy that the rural mass still has a lot of hope and expectation from the government though lot of NGOs are now attracting the people towards their goodwill and work. She finds this as a basic difference between the NGOs and the GOs, that the NGOs are grassroots organisations and try to know the rural problem more closely than government organisations who usually don't try or are not bound to do so.

V. Radha is very happy about the working of the NGOs in Aurangabad division and has a very positive attitude towards their upcoming in developing rural areas. On asked about the Adgaon project of MSSM, she specifically noted that it is one of the best and pioneering projects in the area and the state, which is now being replicated by the government and many other NGOs. She admitted that MSSM has a very good leadership like Shri Vijayanna Borade and if the government avials the support of such leaders in their project, they can change the whole scenario of rural sector. She expects more of the partnership, sharing and collaboration of the government and NGOs for the planning and implementation of development projects.

APPENDIX - III

Questionnaire for Beneficiaries

(On Structured Interview)

1. Name of the beneficiary :
2. Address of the beneficiary :
3. Name of the father/husband :
4. Age with date of birth, if available :
5. Religion :
6. Caste :
7. Sub-caste :
8. Literate/Illiterate :
9. Educated/Uneducated :
10. If yes, Educational details :

11. Family background :
    a No. of family members :
    b No. of dependents :
    c No. of employed members :
    d No. of earning members :
    e Whether joint/nuclear family :

12. Occupation of the beneficiary :
    a Main occupation :
    b Ancillary/secondary occupation :

13. Total annual income of the family :

14. No. of years staying in the village :
15. If migrated, from where :

16. Property particulars :
    a Own house or rented :
    b Type of the house] whether pucca, concrete or katcha :
    c Land :
    d Whether irrigated or not :
    e Agriculture activities :

17. Basic civic amenities and facilities available in the area :
    a School :
    b Hospital :
    c Bus station :
    d Railway station :
    e Electricity :
    f Post office :
    g Community development centre :

18. Do you know about NGO :
19. What do you understand by NGO :
20. Is there any Government Social Welfare Development Works in the area :
21. How many NGOs work in the area :
22. What are their priorities :

23. a Did you receive any benefit from them :
    b If so, from whose was the best :

24. Is there any family income augmentation after receiving the benefits :
25. What are the nature of benefits usually received :
26. Who informs you about these benefits :
27. Who approached you to take the benefits :
28. Do you receive cash or in kind :
29. Do you attend any seminars/workshops :
30. Do you know any of the office bearers personally] NGO :
31. Do you know any Government Officers personally. :
32. Are any of them related to you :
33. Do they visit your house regularly :

34. a Did any of your children get employment/ in any of the NGO/Government Welfare Departements :
    b If so from when :

35. What is the salary they are getting :
36. Whether the salary is sufficient to survive :
37. Before getting the benefits, how you were meeting your daily needs :
38. After getting the benefits, is it sufficient to meet your daily needs :

39. a Are you satisfied with the medical education facilities available in the area :
    b If so, who is serving you better - NGO or Government Department :

40. a Are you satisfied with the Bus services :
    b If so, Private or Government :

41. Are you satisfied with the local transport facilities like auto-rickshaws, cycle-rickshaws, taxis. :
42. What are the main problems you are faced with in your area by way of development/ Government facilities :

43. a Do you have a Gram Panchayat :
    b If so, who is the Mukhiya :

44. Do the Gram Panchayat meet regularly :
45. Do you have a village officer :
46. Have you ever seen him visiting your village/block :
47. Do you have a Mahila Mandal in your village :
48. What work does the Mahila Mandal do ? :
49. Do you have a Yuvak Kendra in your village :
50. What works does it do ? :
51. Do you have Mahilas elected into your Panchayat ? :
52. What are the irrigation facilities done in your area ? :
53. Do you have proper drinking water facility :
54. Has the project improved the water irrigation facilities ? :

55. What is the increase in crop production after the project ? :
56. Has the number of livestock increased ? If so, how much ? :
57. How many animal husbandry are there ? :
58. How many milk dairies are there ? How many litres of milk is supplied to the dairy ? :
59. How many acres of cash crop is grown ? :
60. How many acres of horticulture is grown :
61. Are there any Vocational Training Classes in the village for girls. If so, what. :
62. Are there are any Vocational Training Classes for boys ? If so, what ? :
63. Has the level of education for girls increased ? :
64. Has the age at marriage for girls + boys increased ? :
65. Are marriages arranged with the village or alliance is got from different villages ? :
66. Are people migrating for work to Towns and Cities in summer :
67. What alternative work is done during the dry months ? :
68. How many eligible couples practise contraception ? :
69. Are the children immunized
70. Do you have a Bank in your area - Whether Nationalised/Private/Corporate :
71. What are the development activities you are in need in your area. :
72. What are the other short comings of the present facilities :
73. Any suggestions :

APPENDIX - IV

Questionnaire for Office Bearers of NGO's

(Unstructured Interview)

1. Name of your organisation :
2. Legal status of your organisation :
3. Your position in the organisation :
4. If you are not the Chief functionary, who is your Chief functionary :
5. What are the main aims and objectives of the organisation :
6. What are your resources :
   a. Government grant :
   b. Local contribution and donations :
   c. Foreign contributions :
   d. Income from activities :
   e. Rental income :
   f. Income from other sources - Bank Interest etc. :
7. What is your annual turn over :
8. Break-up of expenditure :
   a. On administration & establishment :
   b. On maintenance and up-keep :
   c. On aims and objectives :
   - Population :
   - Poverty Eradication :
   - Education :

- Medical :
- Relief and Rehabilitation :
- Agriculture :
- Natural Calamities :
- Other social activities :

9. Do you file your returns/statement of accounts regularly with the respective departments.
10. Do you have a list of beneficiaries :
11. Do you receive Government grants, if so through whom, for what purpose ?
12. Do you receive foreign grants, if so, what are the source of connection ?
13. Total members of your Organisation:
14. Total govering/Executive Committee members of your organisation
15. Do you have elections every year, if not how often ?
16. In this organisation based on religion Secular or inter religious
17. If religious, which religion does it belong to :
18. If you are availing Govt. grants, do you experience any difficulty in obtaining such grants
19. Did the Govt. officials come to you with the information of available grant, or did they encourage you in taking such grants
20. Which departments does your organisation usually approach for such grants
21. Did the Govt. officials visit your organisation for inspection and are they satisfied with your work
22. Are the beneficiaries satisfied with you work on social development
23. What are the activities done during the last three years by your organisation] say in brief
24. Do you feel these activities are sufficient for the purpose of serving the people of your area. If not, do you have any other suggestions.
25. a Do you have a similar NGO in your area working the similar aims and objectives b If so, how many are three ?
26. Does the Govt. Dept. also extend the same programme/projects which you are undertaking in the same area
27. Do such activities have a positive or negative effect
28. How do you regulate your indicators and what are those
29. Do you have a monitoring system or evaluating system in you area
30. Do you agree that there is more scope for further activities
31. According to you what are the developmental thrust one should consider to be carried out in this area.
32. You specific suggestion if any towards a new developmental vision.

www.ingramcontent.com/pod-product-compliance
Ingram Content Group UK Ltd.
Pitfield, Milton Keynes, MK11 3LW, UK
UKHW021450280726
14060UKWH00001BA/334